I0483679

Linking Mind and Matter

A New Perspective for a Scientific Model

That Unifies Nonlocal Mind and Local Matter

RW Boyer

Institute for Advanced Research Malibu CA

First edition 2012

Jacket by Liz Howard Graphics

Library of Congress Number: 2012908733

ISBN: 978-1-4774-6061-0

CONTENTS

Introduction

A *whole new perspective on scientific theories linking mind and matter is introduced in non-technical language. From physical to quantum gravity and unified field theories, how key dilemmas are resolved through levels of nature including nonlocal mind subtler than the physical is overviewed. Ancient holistic views such as the Vedic model are shown to be the direction of scientific progress toward unified field theory that integrates objective and subjective levels. Epistemological concerns about validation are discussed in the conclusion.*

This concise book overviews progress in modern scientific theories 'connecting the dots' toward a unified understanding and experience of nature. It shows how linking mind and matter requires ontologically real levels of nature beyond the physical.

First, progress from classical Newtonian and relativistic physical theories to quantum and quantum gravity theories is summarized in simple terms. This includes orthodox, many-worlds, objective reduction, and neorealist interpretations of quantum theory. Then string theory, loop quantum gravity theory, big bang theory and emerging alternatives are overviewed toward unified field theory.

Introductory descriptions of these theories are included to highlight the essential steps of progress and make them accessible to a wider range of readers. The descriptions also are helpful for understanding how the progress is toward the ancient holistic view that has been overshadowed by an engrained reductionism. The theories are examined from the uniquely revealing perspective of how mind fits into them, and tacit assumptions are made explicit.

With subtle insights into cutting edge theories, and a heuristic of *levels of spacetime* to thread them together, an integrated model of subjective mind and objective matter is described. For the first time in modern science, a logically consistent picture is presented of how your arm for example actually can be guided by your mind.

Modern physics progressed as an *objectified* reductive investigation, for the most part leaving out subjective mind. This led to a model of nature with basically one ontological level: the causally closed objective physical world, sometimes called *materialistic monism* or *physicalist realism*. More recently a two-level model adding unified field theory has been developing, which as we will see is turning everything downside up. But quantum theory required consideration of how to transition from quantum wave functions in imaginary mathematical space to real physical space; and this brought subjective mind into the picture. This issue is revealing that a two-level model isn't ontologically rich enough to have room for real mind.

Recent models by prominent theorists David Bohm, Roger Penrose, Henry Stapp, and John Hagelin are examples of three-level models. Deep insights into physical and quantum theories support the main thesis in Part I that these models are converging toward the holistic *three-in-one* model in ancient Vedic literature. This model includes how ontologically real mind fits into the picture.

The main thesis in Part II is that the Vedic model has considerable detail about linking mind and matter. It is particularly relevant to theories developing a new picture of how to address the old mind-body problem via levels of spacetime beyond the physical.

But the ontologically richer models emerging in modern science are increasingly difficult to validate, even with our most advanced experimental methods. This has led to speculative mathematical models not grounded in empiricism; and concerns are arising about the apparent trend in physics of *faith* in mathematics. It calls for a reconsideration of scientific epistemology in order to investigate nature in an expanded ontology beyond the physical. The main thesis in Part III is that the Vedic tradition applies systematic direct empirical means of validation that complement indirect objective experimental means. The book concludes with a brief discussion of epistemology in Yoga and its subtler approach to empirical validation.

Each of the topics in this short book is included to clarify the coherent basis for explaining how mind links with matter. An immensely significant and fortunate step is the current paradigm shift that unifies reductionism and holism.

PART I

LEVELS OF NATURE

Chapter 1

Physical Theory

Newtonian classical physics identified matter, forces, and space and time as fundamental constituents of nature. There were an unknown number of matter types and three forces (gravity, electricity, magnetism) in a non-changing *empty* background of infinite space and eternal time with a universal *now*. Newton believed mind and God also existed, but couldn't account for their relationship to the fundamental constituents. Matter and mind, objective and subjective, as well as physical and spiritual, were not linked together.

The perspective that the universe is a cosmic machine eventually included the human body, [1] and soon after the mind. [2] This perspective set the stage for reductive physical theory that has been the mainstream scientific worldview. The assumption that physical objects, with different degrees of solidity and tangible presence, are the only objects that exist in nature is a basic tenet of physical theory. It is drawn from the belief based on ordinary sensory experience that objects we can hear, touch, see, taste, and smell are the only ontologically real objects, existing objectively independent of our subjective minds.

This might seem to suggest that the mind is something other than the physical. But physical theory also assumed that the mind is in the physical. The mind-body problem and the link between subjective and objective were presumed to be resolved, because the mind eventually will be completely accounted for by physical brain processes. This reduced the focus of scientific research to the search for the essence of matter, which eventually would account for mind— if correct.

Four forces and local causality. Reductive investigations analyzed matter to atomic, sub-atomic, nuclear, and sub-nuclear particles. All motion involved matter particles bumping into each other—the 'billiard ball' model of causality. Eventually it was recognized that this model couldn't explain phenomena such as what holds sub-atomic particles into atoms, and how some atoms fall apart by radiating sub-atomic particles. Two other forces were proposed: the strong nuclear force holding atoms together, and the weak nuclear force allowing atoms to emit particles. Also, electricity and magnetism were recognized to be unified as one force. Four fundamental forces were then identified in a *local particle interaction* model of causality that was believed to account for everything that exists as real in our 'visible' universe: gravity, strong nuclear, weak nuclear, and electromagnetic.

The local particle interaction model of causality also was used to describe how the forces influence particles. The forces were modeled as functioning via transient or *virtual* exchange particles that mediate change between matter particles through absorption and emission.

A core feature of this theory is that motion cannot be faster than the photon light particle. The theory accounted for most all change in the physical world, from atomic to sub-atomic particles to galaxies and everything in between. We still don't know what particles and forces are, but calculations of their behavior are extremely accurate.

Relativistic physical theory

Einstein advanced physical theory dramatically by linking gravity with space and time into *spacetime*. Space and time were no longer understood to be a non-changing infinite eternal background, but relative to each other. The gravitational force was the curvature of spacetime, not a separate force functioning *in* it. It was said to be the natural result of the interactions of tangible 'material bodies.'

Spacetime was described as a smooth continuous geometry of one temporal and three spatial dimensions. Curvature of spacetime due to the presence of matter/energy generates the gravitational force. Extreme curvature can produce black holes, in which nothing escapes the immense pull of these dense centers of gravitational attraction.

The observer is part of spacetime. According to relativistic physical theory, there is nothing outside of the spacetime gravitational field. Spacetime can expand, but seemingly reasonable questions about what exists outside of it—such as what it expands into, or possibly shrinks from—were considered meaningless because there is no background. Also in this relativistic view, there is no universal *now*.

Another core feature of the relativistic spacetime gravitational field concerns what happens when observers like us move toward or away from each other. Observers with similar motion will have basically the same results in measuring time and distance, but differences show up if relative motion were to become extremely different. This means that the space and time we ordinarily think we exist in is not a separate background. Our physical bodies are part of this field, and even how old or wide we are changes compared to other objects, more noticeable as relative motion nears light-speed.

In Einstein's relativistic theory light-speed and the frame of reference of the observer define the *light cone*. Because nothing can travel faster, there is no possibility of one object or event causally affecting another outside the range of light-speed. The past light cone includes everything that could have influenced the particular observer from the past, and the observer's future light cone expands at light-speed the range of potential causal influences into the future.

This describes *local causality*, the view that all causal relations take place by matter particles interacting with each other within light-speed. However, relativistic physical theory didn't account for what initiates the causal chain, how to include mind if it is real, and other related key issues to be made explicit as we proceed.

Perhaps a more tangible way to view relativity theory is to picture the spacetime gravitational field as a medium—sort of a gigantic glob of spacetime plasma. All objects in the physical universe exist as part of it, and change according to how they move relative to each other. This is more integrated than Newton's theory because space and time are not a separate background from material bodies. But the theory at least tacitly assumed that everything is physical—even all parts of us including our minds. This contributed to serious questioning and widespread doubt whether mind, as well as God, actually exists.

Relativity theory was the most successful and comprehensive theory in science. However, the most significant implication of the theory seems not generally appreciated. It is only now starting to be revealed with deeper levels of unification, such as with supergravity theories [3] in which all expressions of physical matter can be said to emerge from the spacetime gravitational field.

8

The spacetime gravitational field can be viewed as an ethereal substance or *matterstuff*. This is the essential point of the finding in relativity theory that space and time are not a separate background. It is not that spacetime results from the motion of matter. All matter is built of the relativistic spacetime gravitational field—like an ocean of water with icebergs in it made of water. In other words, relativistic spacetime is the matter. This subtler view is crucial to the more integrated understanding relativity theory gave us. It will be an important emergent model of levels of nature as we progress beyond the physical to subtler levels of spacetime. For practical purposes in our ordinary daily life, however, it is as if relativistic spacetime is a separate background, as in Newtonian physical theory.

Levels within spacetime. Nature is now viewed as structured in layers from macroscopic to microscopic atomic, nuclear, and sub-nuclear. Objective investigation has involved reductive probing of smaller time and distance scales and higher energy and temperature states, in search of the essence of matter. The range of scales can be characterized simply as follows:

Ultramacroscopic levels: Infinity to cosmic expanse
Macroscopic levels: cosmic expanse to $\sim 10^{-3}$ cm
Microscopic levels: $\sim 10^{-3}$ cm to $\sim 10^{-8}$ cm
Ultramicroscopic levels: $\sim 10^{-8}$ cm to $\sim 10^{-33}$ cm
Unified field level: $\sim 10^{-33}$ cm or infinitesimal to Infinity

The resolving power of our ordinary senses for direct sensory observation is in the comparatively limited macroscopic range. The wavelength of visible light for example is in the range of 10^{-4} cm, too wide to observe directly anything smaller than a cell. Visual observation has been extended with the aid of probing tools such as the electron microscope to about 10^{-8} cm, still larger than an atomic nucleus. In other words, no one yet has directly seen an atom in detail, but indirect evidence for it is considered extensive. As we will consider, however, this view of layers of nature is inadequate, and an additional metric of levels of interdependence, entanglement, and holism as fundamental to the structure of nature is emerging.

The process of observing influences the observed. With more powerful indirect methods of probing nature such as particle accelerators, research has gone far beyond evidence obtained directly through the ordinary senses. The results of these indirect methods are macroscopic phenomena observed via the ordinary senses that are predicted by and dependent on conceptual models of events theorized to occur at smaller unobservable scales.

But at tiny ultramicroscopic scales, probing and measuring tools are thought to influence and even significantly alter the theorized objects being investigated. This makes objective experimental investigation of independent objects much more challenging on these theorized finer-grained layers of nature.

Analyses of theorized objects at unobservable scales increasingly rely on conceptions about what is being measured and what the process of measurement means. Functionally deeper cognitive processes of logical reasoning are relied upon more than sensory perception; and it is increasingly recognized that what is observed depends on subjective processes in the observers. This is evident in the current prominent role of mathematical models based on logical reasoning in formulating and even evaluating theorized events beyond ordinary direct sensory observation, given the increasing difficulties of reductive experimental research at the limits of the physical universe for both extremely large and extremely small time and distance scales.

A major change from classical physical theory to quantum theory is that these issues were recognized as core to the *measurement problem* and the influence of the observer in *creating* what is observed. Heretofore tacit assumptions about the object of investigation, the probe, evidence of their interaction, as well as the role of the observer now needed to be examined explicitly in investigating the essence of matter.

Historically such subjective issues were the purview of philosophy—and to some degree also religion—associated with the *mind-body problem* and more recently the *hard problem* of consciousness. These issues are now prominent in physics, as they have been in psychology and neuroscience associated with the *explanatory gap* between consciousness, mind, and the material brain.

Emergence theories and levels of nature

Reductive physical theory relates to what is sometimes called epistemological emergence. This refers to emergent properties being *reducible to* and caused entirely by their parts or bottom-line physical constituents—even though it may be quite hard to derive and predict them. [4] This is consistent with the reductive physicalist worldview still prominent in modern science, despite strong implications in quantum theory that it is untenable.

In this meaning of emergence, the emergent properties *do not* have real causal power over the parts from which they emerge, being entirely due to the underlying parts. This is consistent with mind and consciousness as fundamentally non-existent and having no causal power, such as in eliminativist theories in cognitive science and modern science generally. It means that our conscious minds are not real, we have no free will, and we are just zombies or robots.

A primary body of evidence to support this view is that cognitive functions are lost when their corresponding neural correlates are damaged or anesthetized. But the evidence has been over-extrapolated to conclude that higher-order functions are thus products of lower-order physical processes. It doesn't eliminate the possibility that the connection between them is broken and higher-order functions are just no longer accessible. This is exemplified in the well-known analogy that destroying a radio doesn't destroy the transmitter and signal, though the signal can no longer be heard via the radio. This is the direction of cutting edge research, which has not found mind in the physical brain, and is now exploring deeper quantum field explanations to account for mind and consciousness.

In recent years an interesting but somewhat surprising alternative to epistemological emergence sometimes called ontological emergence has become widely popular in scientific accounts. It refers to properties emerging into a holistic system that is *not reducible to* its underlying parts or their relationships, but apparently *does* have causal power to direct the underlying parts. [4] In this meaning of emergence, the whole emerging from the parts is more than the sum of the parts, and somehow is able to insert into the closed physical causal chain super-ordinate control of the lower-order physical parts.

But the emergent whole is not attributed independent real existence separate from the underlying physical parts that are said to generate it. In this view the higher-order whole of unitary conscious mind has just epiphenomenal status. It thus seems to be a misnomer, in the sense that the emergent ontology is not real. It is an attempt to account for a real role for conscious mind, within the reductive physicalist worldview that is inconsistent with such a role.

The causal power of higher-order conscious mind on lower-order parts is logically inconsistent in this model, because the causal chain that creates the emergent whole has no breaks in it to insert a new causal influence that could control the underlying parts. It is not ontologically real emergence, in that no new and real level of nature where the whole might exist is created. Rather, it is *epiphenomenal* emergence, and as such its causal power is also not real but epiphenomenal. In this second meaning of emergence, we again have no real mind and no free will, remaining just zombies or robots—only material bodies, within materialistic monism or physicalist realism.

The whole can be said to be greater than the sum of its parts in the sense that new capabilities and behaviors emerge when the parts unify into a whole. But with respect to physical processes, the whole is not greater than the sum of its parts in the sense that a complex collection of interconnected parts *creates* a higher-order ontologically real level of nature in addition to the physical that could have top-down power to direct unitary behavior of physical parts. New physical effects emerge from new combinations of physical objects and processes, but not with real causal power over the closed physical chain of events that connects the parts. With respect to brain function, neural complexity allows for new behaviors; but no new ontologically real level of nature with causal power over the brain and behavior emerges inside or outside of neural structures in the brain.

For concrete analogies, the conception of several mountains as a mountain range doesn't empower the mountain range to control each mountain in it. A forest similarly doesn't control the trees in it, though the collection has ecological effects in addition to each tree. Four hands from two different individuals may allow greater breadth and efficiency of task performance, but don't automatically create a third mind. Correspondingly, a network of computers can increase capacity for problem solving, but no super-ordinate conscious mind—physical or non-physical—is created. Importantly, in this view the Internet is not, and does not create, an ontologically real mind.

Again, the whole as a collection of parts does generate new phenomena. But in the physical world, the collected whole influences the parts via the same mechanisms through which the parts influence each other. For a global influence of the whole directly on each part, or for each part to influence directly the whole, physical connections between them are necessary in a causally closed physical network.

Complexity increases rapidly with more interactions between the parts of an integrated physical system. At some point smaller time and distance scales or finer-graining of lower-order parts is needed to implement interaction between all parts of the system. This is the current direction in building microstructures that allow more powerful computational functions in information science, such as nano-technologies, potentially enhanced further by more interconnected holographic systems. But causally efficacious conscious minds don't magically appear with increasing physical complexity and interaction at smaller time and distance scales. If they did, they would need to exist as real somewhere beyond the physical.

As quantum theories now to be overviewed have developed, models are being proposed of levels beyond the physical that include nonlocal quantum mind which logically *could* cause changes in the physical causal chain. But this requires that the physical causal chain be *open*. It just doesn't look open from the coarser-grained physical perspective. It suggests more fundamental, real levels of nature.

This relates to another meaning of emergence, consistent with the holistic unified field-based perspective as the source or bottom line of everything in nature, discussed later. [5] In this meaning, the whole precedes the parts, and the parts emerge as limitations of the whole. Instead of the reductive physicalist view that consciousness and mind are non-existent or just epiphenomenal, they are the basis of matter and shape its expressions all along. In this view non-physical mind underlies physical parts and subtly guides them to create biological organisms that intentionally direct their own unitary behavior.

This meaning of emergence relates to the *consciousness-mind-matter ontology*, basically opposite of the matter-mind-consciousness ontology in which conscious mind is said to be a product of the physical brain, turning everything downside up. It is the only theory of emergence that can logically support the causal power of mental intentions. And it means that the physical causal nexus is not completely closed as it appears on surface levels. In further contrast, in the other emergence theories described above change is *fundamentally* random, which also disallows free will because ultimately the closed causal chain is initiated randomly.

This more holistic unified field-based understanding of emergence refers to the expression of capabilities already present in the whole. Again, the whole creates the parts. The ultimate wholeness is the unified field as the bottom-line source of order in nature. This type of *phenomenal* emergence will become clearer as we progress in our examination of quantum theories and how conscious mind relates to them. It is more consistent with the ancient Platonic view, as well as the ancient holistic Vedic model to be discussed once progress beyond the physical has been overviewed.

After Einstein's tremendous success with relativity theory and related discovery of the equivalence of energy and mass ($E=mc^2$) resulting in a more integrated scientific picture, he took on the task to build a unified field theory. Though spending much of his later years on this project, he was not able to complete it. As we will see, this is due to trying to unify everything within the physicalist view of a closed physical chain of causality, which is an inadequate ontology.

We will now overview progress beyond Einstein's monumental accomplishments to levels of nature in addition to the physical. These more advanced but more abstract theories extend into an underlying background to the relativistic spacetime gravitational field. In other words, relativistic spacetime emerges from an even subtler substance, medium, or field of nature. These theories include more expansive conceptions of spacetime, in which mind with thoughts and feelings really does exist, toward a coherent theory of the infinite space and eternal time of the unified field. This is a much bigger picture.

Chapter 2

Quantum Theory

A huge step toward the bigger picture than relativistic physical theory is quantum theory. Even more successful and integrative, quantum theory has held up in all experimental tests of its predictions including ones classical physical theory and relativity theory were unable to address.

Einstein made major contributions to quantum theory, while also having major concerns about some conclusions drawn from it. These concerns relate not only to matter but to mind. In physical theory and modern science generally, mind was viewed as separate from matter (mind-matter duality, subject-object independence), but also was presumed to be nothing other than matter. This is inconsistent, and quantum theory severely challenged it and is rendering it untenable.

A background to ordinary spacetime. The concept of a quantum further integrates particles and forces in an abstract mathematical field. Quantum fields are described as vibrating at discrete potential energy states, forming *quanta* or wave packet potentials. Matter particles are viewed as having no material existence other than being potential stable states of abstract quantum force fields.

In other words, *matter doesn't have a material basis.* This can be interpreted to mean that the matterstuff of the relativistic spacetime gravitational field is not the bottom line of nature. It emerges from a more fundamental non-material substrate; thus it is *background dependent*, rather than background independent as in relativity theory.

As core conceptions in quantum theory are unpacked, however, we will see that there has been considerable difficulty getting beyond physical theory to accept a real background to matter, even though implied by quantum theory. In other terms, bridging from quantum potentials in imaginary mathematical space to the ontologically real world of physical spacetime has been quite a major challenge. Recent quantum gravity theories that can be viewed as a bit closer to crossing this bridge will be discussed in the next chapter.

In quantum field theory particles are described as potential fluctuations of abstract fields that vibrate at multiples of the *Planck scale*, named after Max Planck who in 1899 took a first major step toward quantum theory. He derived a universal unit of measurement from actual experiments, based on light-speed, Newton's gravitational constant, and Planck's constant. The Planck scale is the incredibly tiny distance of 10^{-33} cm (a millionth of a billionth of a billionth of a billionth of a centimeter), the incredibly brief time of 10^{-43} seconds (the time it takes a photon light particle to cross the Planck length), and the incredibly powerful energy of 1019 GeV (gigaelectron volts).

The Planck scale is held to be the smallest possible size of any object and any measurement in nature. The mathematical notion of dividing distance to a dimensionless point doesn't quite seem to work in the physical world once we get down to the smallest scale. At the Planck scale, our ordinary notions of time and distance break down. An important thread as our overview proceeds is that a breakdown of ordinary notions of spacetime doesn't invalidate all notions of it.

In quantum field theory the patterns of potential wave vibration determine the quantum field's role as either a particle or a force. Stable potential states of vibration are associated with the *matter particle* quality of the field (usually fermions). Transient potential states of vibration are associated with the exchange particle or *force particle* quality of the field (usually bosons). In quantum field theory a force is an effect on a stable matter particle by a transient exchange particle that passes between matter particles. All motion and change are described as the exchange of these quantum wave packet potentials in a particle interaction model, still referred to as particles but no longer really thought of as solid matter. Though it sounds like quanta are real objects, their ontological status as real was not part of the initial orthodox interpretation of quantum theory, discussed soon.

Quantum fields are also described as capable of being in a least excited or ground state, the *vacuum state*; but this is not absolute zero energy with no fluctuations or vibrations. The quantum field is said to exhibit zero point motion or quantum vacuum fluctuations whether in its particle, force, or vacuum state. This vacuum energy is attributed to be the *inherent dynamism* of quantum fields.

Quantum field theory represents an object as a *superposition* or overlapping of all possible shapes of the quantum field. Each potential physical object is a probability wave or cloud of *tendencies to exist* that translates into real physical states with a measurement or observation. In this model, unbounded mathematical wave potentials somehow appear as bounded particles. This suggests a more wave-like than local particle interaction-like causal model, discussed later.

In quantum field theory the gravitational field is sometimes envisioned as a soupy froth of transient particles coming in and out of existence instantaneously—*spacetime foam*. This importantly means that space is not empty. It also suggests that change is initiated from inherent random fluctuations of quantum fields, also discussed later.

Observer/background relationship? In quantum field theory it has been theorized as possible for a physical object on one side of a physical wall to appear on the other side without traveling through it in the ordinary way. This *quantum mechanical tunneling* is described as a common process associated with very short distances such as nuclear reactions in radioactive decay. It has been speculated further that it may be possible to 'travel' anywhere through these quantum mechanical tunnels—*wormholes*—with instantaneous connections between regions of the universe, even outside the light cone. This 'travel' would seem to imply bypassing limitations of the relativistic spacetime gravitational field. And it further implies that quantum fields in some way may be real fields, not just mathematical concepts.

Theories are being developed to describe how the information about a physical object to 'port' it could be extracted from the object—which apparently would decompose it. It is not clear where the information about 'you' for example would exist as you are being decomposed, how it could instantaneously appear at the *intentionally selected* new location, or how then to recompose it back into you.

The point here is that porting to parts of the universe outside the light cone contrasts with the defining features of the spacetime gravitational field. Consistent with many peoples' intuitions, it suggests that space is not empty, there may be a background with a universal *now*, and even there may be some way to direct intentionally where to go via this background—a possible causal role of mind.

A related major challenge is that in quantum theory there isn't seamless motion of objects from one place to another as conceptualized in physical theory. Distance between two physical locations has only a statistical meaning based on repeated measurements, with at least some uncertainty. This is said to be due to quantum indeterminism and the inherent dynamism of random fluctuations at the Planck scale—again, held to be the smallest possible size and the smallest possible change in space and time.

Ordinary measurements of sensory objects are basically thought to be averaging across potential random fluctuations, with the potential fluctuations for practical purposes canceling each other out. Calculations based on quantum theory end up yielding the same results as at the much larger, coarser-grained time and distance scales of classical physics research.

But Einstein didn't accept fundamental randomness or indeterminacy as the very heart of nature, which he felt challenged the central pillar of deterministic causal relations in science. He expressed this concern in his famous comment, "I cannot believe that God plays dice with the universe (as quoted in Herbert, 1985)."[6]

Even more significant challenges confronted physical theory with respect to mind. Again, a basic tenet of classical science is that objects in nature exist independent of the observer, associated with object-subject independence and believed by many including Einstein to be fundamental to science. But also, in physical theory the mind was presumed to be completely accounted for by physical processes in the brain. Minds were nothing more than, and entirely reducible to, material bodies. Despite massive amounts of research, however, the mind has not been found in the brain/body. If it exists in the brain/body only, then it must be in the closed causal nexus of the physical. However, there is no room in the closed causal chain that started long ago to unlink and insert a new causal influence, such as a mental intention, to have a real causal effect in nature.

And again, the logical conclusion in physical theory is that mind is at best epiphenomenal with no causal power. Quantum theory required serious reconsideration of this view, as noted earlier, which is continuing to take place and growing in recognition.

Orthodox interpretation

In classical physics, particles are represented mathematically as dimensionless points with *no extension* in space. But as quantum wave potentials, they have *infinite extension* in mathematical space. Understandably, how unbounded wave potentials in imaginary mathematical space transition into real objects in physical space has been a great mystery in quantum theory. This brought the issue of mind directly into the picture. It has led to a revival of the mind-body problem and reconsideration of the ontological status of mind and consciousness as major issues even in the field of physics—which historically didn't address mind at all due to the intention to be *objective* and the assumption that mind is nothing other than matter. But quantum theories have been progressing to the recognition that matter is not the bottom line of nature, and that mind somehow is involved in creating the phenomena of real matter.

The initial orthodox Copenhagen interpretation recognized these issues, but didn't quite address them. It asserted that quantum theory is a model of the relationships between real objects in ordinary physical reality and mathematical quantum concepts to understand them; it is a recipe for how to explain some phenomena not accounted for in classical physics. But it could not be about *quantum reality*, because such a reality was still believed not to exist. [2]

Accordingly it is fantasy to think that an individual electron is heading toward a TV screen to create a blip on the screen. Before the blip, the electron has no reality other than mathematical probabilities of existing at certain times and places when measured. If it is imagined as existing and traveling before observing or measuring it, it would be as if it were traveling everywhere, including forward and backward in time—quite different from the physical view of objects.

Observer creates observed? To attempt an explanation of the transition from imaginary mathematical quantum wave potentials to discrete real physical objects, it was posited that the quantum wave function *collapses instantaneously* with a measurement or observation. This was an enormous and baffling change in physics, because the process of observing in this view necessarily brought the conscious mind of an observer to center stage of the picture.

As to how and where quantum wave function collapse occurs upon observation, and what actually collapses, the initial orthodox Copenhagen interpretation built an 'inviolable wall' beyond which it was argued that further answers about the collapse are not possible. It was reasoned that this is because nature has no more information: beyond the 'inviolable wall,' presumably nature is indeterminate and fundamentally random.

This view practically divides nature into probabilistic *tendencies to exist* and discrete real physical objects that emerge only in the process of observing or measurement. It reflects the gap between the sensory world of physics and the conceptual world of mathematics, and the difficulty but also necessity of bridging the gap. The 'inviolable wall' notion seems inconsistent with the fundamental objective of science to develop rational, systematic accounts of how processes take place in nature. It is odd for a scientific theory to assert that no further explanation is possible. The model of quantum wave function collapse needs to be unpacked in order to get to the key issues it implies about nature, and especially about the newly recognized role in physics of conscious mind. A comprehensive theory needs to account for both subjective and objective, as well as how they interact. We thus then had a quantum wave function in imaginary mathematical space, an objective physical world of real objects, and the process of observing that connects them somehow. But how?

Some theorists suggested that the act of making a record—a film recording, needle mark on a scroll, or any automated scoring system—counts as an observation. This view asserts that wave function collapse can occur with simply a measuring device making the observation or measurement, so there is no need to complicate the picture by bringing in the subjective mind of the observer.

But would binoculars for observing nature that are just sitting on a table, a ruler for measuring distance, a clock for measuring time, or a thermostat for measuring temperature *create* or be accompanied by wave function collapses on their own without a conscious observer involved? If any event can be said to be measured or observed even if an observer is not present and not involved, then how come all events apparently occurring in nature don't cause wave function collapse?

An inert physical object, even if given the role as an observing or measuring device, presumably is not conscious and cannot sense anything. And as a classical object its wave function already would be collapsed, so there would be no 'uncollapsed' wave function in it that could possibly collapse into another classical object via observation.

Indeed, then what and where is the quantum wave function that collapses to get a discrete physical real object? Does it exist as ontologically real in nature apart from the observer, or is it just in the observer's mind as an imaginary mathematical concept? Is there a real conscious mind to begin with inside the physical observer? These fundamental questions need answers for a rational model of quantum wave function collapse, but it seems were not addressed.

In trying to identify the crucial point where the theorized wave function collapse might occur, mathematician and physicist John von Neumann could not identify any particular place outside of the observer's conscious mind. He reluctantly concluded that it would have to occur where the chain of quantum wave functions ultimately stops. That is the observer's consciousness. [7]

For example an observer measures the amplitude of an acoustic signal using a dB meter to be 80 dB. The dB meter is a measuring device for making acoustic measurements, but is not in itself a conscious entity. It might be said that the observation occurred when the sound waves contacted the dB meter, or the indicator pointed to 80 dB, or light reflected off the indicator into the observer's eye, or the retina was activated, or the optic nerve was activated, or the visual cortex was activated, and so on. All these events are describable theoretically as probability wave functions. No matter how far this chain is traced, until conscious mind enters the picture there is no observer, observed, or process of observing. [8]

In this orthodox interpretation, the collapse must take place to get physical objects, and there was no other logical place for it than the conscious mind of the observer. This identifies the crucial place of quantum wave function collapse as consciousness. [6] It is remarkable that this interpretation seemed for the first time to place subjective conscious mind as essential to objective science. It might even be viewed as implying that conscious mind has an actual role in the casual chain of events, requiring an even bigger picture with additional real levels of nature including real minds to account for wave function collapse and a causal influence of mind on matter.

But again, it was assumed that conscious mind, if it exists at all, must be part of the classical physical world, and thus inaccessible to be examined and modeled using quantum principles. In orthodox quantum theory at least so far, there is no way to formulate a wave function that includes the classical observer.

Relativistic physical theory also assumed that the observer is a classical physical object. But the observer's conscious experience is not accounted for in either quantum or classical theories. In both theories the observer views nature from outside, while also generally rejecting the notion of something outside. Orthodox quantum theory didn't account for conscious mind, but paradoxically gave it a unique role not given to anything else in nature.

This leads to another major dilemma: how can any quantum wave function collapse occur to get a classical observer when the observer is a prerequisite for the collapse? How could a discrete physical brain, necessary for conscious mind, exist before conscious minds evolved in nature? It is inconsistent to hold that classical objectivity requires conscious subjectivity for quantum wave function collapse and also hold that classical objective processes occurred long before conscious minds existed. This suggests a much bigger picture, as Einstein and orthodox proponents might also have intuited. The bigger picture would explicitly address how conscious mind emerged in biological evolution, necessary for orthodox interpretations to be consistent.

Many-worlds interpretation

The conclusion that objective science must include subjective conscious mind was quite unsettling, for one reason because it seemed to require 'soft' subjective topics as crucial to 'hard' objective science. One alternative, the *many-worlds* interpretation associated with mathematician Hugh Everett, [9] attempted to reestablish objectivity by eliminating entirely the concept of wave function collapse. But it did this by eliminating, or at least ignoring, physical reality and the known natural laws of ontologically real physics.

Nature as mind-worlds? Rather than collapse of a mathematical quantum wave function, in this 'no-collapse' model each mathematical possibility branches into its own parallel world with its own mind. This resolved the measurement problem by eliminating wave function collapse and the role of the conscious observer in it.

For example, shuffling a deck of cards creates a new parallel world from each possible outcome. It is sometimes even radically interpreted as suggesting that there are an unlimited number of copies of you created from each possible outcome of every event.

But this brings mind even more into the picture. It requires new parallel worlds, each with its own mind, appearing automatically, instantaneously, out of nothing, with no mechanics for their creation. Rather than actual physical parallel worlds—which would at least seem to violate the law of energy conservation—apparently the worlds are *mind-worlds* in imaginary mathematical space. But it doesn't address how any real physical world comes about, or how the observer experiences his or her own world as a discrete world with real energy and mass if the quantum wave function never collapsed.

And importantly, it doesn't account for how the sense of continuous identity or 'I', as well as any successive logically consistent events, occurs in the branching of worlds. There is no information that ties one world to the next in the branching process that could maintain continuity of memory, no way to experience other worlds, and not even the same minds across theorized events and worlds.

This fragmenting view generates more questions about mind and matter than it answers. It is another example of a mathematical conjecture that didn't seem to consider ontological issues of what is real in nature. It didn't bridge abstract mathematical possibilities and physical actualities, a continuing issue as we progress in this review.

Objective reduction interpretation

The next interpretation, *objective reduction*, is a completely different approach to eliminate the vague subjective concept of quantum wave function collapse via observation and get the observer out of the picture. In this interpretation there is a collapse, but it doesn't involve an observer or process of observing. [6] It appropriately takes subjective mind out of the picture with respect to quantum wave collapse. But it also appropriately brings it back into a bigger picture by recognizing continuity of the observer's experience across events, which the many-worlds interpretation didn't do.

A real quantum background. A prominent version of the objective reduction interpretation is the *consistent histories* approach associated with *decoherence.* [10] In this version quantum wave collapse into a classical object takes place spontaneously through interaction with the physical environment. The wave-like nature of objects can be inferred from quantum wave interference effects when the wave pattern is coherent and not disrupted by environmental influences. But interactions with the environment create a *decoherent* effect that breaks down the interference pattern.

Thus wave function collapse doesn't involve a conscious observer, isn't instantaneous, and occurs in what is considered to be real time and space, at least in that it interacts with real objects. For example, a potential grain of dust interacting with molecules, sunlight, microwaves, and so on will decohere in a tiny fraction of a second. [10]

What is particularly significant about this interpretation is that it asserts that quantum and classical objects causally interact. Up to this point, quantum waves were imaginary mathematical functions, not real objects. Objective reduction implies that quantum waves are ontologically real—suggestive of *quantum reality*. It would seem they must exist on their own as real, if observers are not involved at all.

In this interpretation causal models of matter particles bumping into or interacting via absorption and emission are inadequate. Real quantum *waves* interact with real physical particles, and their wave functions collapse due to the interactions, independent of an observer. The many practical applications of quantum mechanics in recent decades also support a real quantum level of nature.

In the objective reduction interpretation, quantum waves are no longer just concepts in imaginary mathematical space but are now theorized to interact causally with physical matter in our real world. And as we will now consider, it also means that quantum theory— and a rational science generally—must be logically consistent with respect to the empirical, experiential perspective of observers.

Nature must be orderly to the observer. Decoherence is not sufficient alone to narrow down different observer perspectives to a consistent discrete physical state. Another way of saying this is that decoherence in the objective reduction interpretation doesn't offer a basis for how questions about the world requiring observations or measurements correspond to answers that are consistent with the questions. The *consistent histories* framework can be said to acknowledge this issue, which the previous interpretations didn't do. It not only emphasizes context dependence for decoherence, but also logically consistent histories from observer perspectives across events.

The context serves as a selection process that spontaneously narrows down quantum possibilities, independent of a conscious observer. But from the perspective of an observer, these processes must be consistent across events in time for a discrete outcome. This recognizes consistent experience of observers as necessary for an *empirical* science like physics—even if not for wave function collapse.

As a simple analogy, a sentence is constructed in a consistent series of words with the first word setting a context that narrows down possibilities for the next word. If the first word is 'the,' then the next word likely would be consistent with it, such as 'tree' rather than the unlikely 'however.' For another analogy, in baseball, for a player on first base the next consistent location is most often second base, depending on the context in the game.

In this view there is one big world with many minds and observer perspectives in it. The world we get from observation depends on the questions we ask, as well as the measurement choices we make and the measurement processes we set up and use. Definite answers emerge from questions in a logical, context dependent manner. [7]

Further, the emphasis on context dependence and consistency through change ultimately gets back to the *initial conditions* shaping the universe. The entire history of the universe that gives consistent answers to measurement questions needs to be consistent with the initial conditions—in order for a natural world to emerge that can be known rationally through scientific observation. These principles suggest that the initial conditions must include all possible consistent answers to questions about nature.

Moreover, the initial universe must have been orderly, with low entropy, which has huge implications for science. This relates directly to Einstein's opposition to the view in the orthodox quantum theory interpretation that nature is fundamentally random. Change was assumed to be due to random quantum fluctuations; but it needs to be logically consistent for observers of it. At least very early in the process of creation, change needs to be non-random and orderly—if not earlier, as we will discuss when we consider holistic unified field theory and its relationship to conscious mind in upcoming chapters.

The principles of decoherence and consistent histories also are consistent with the asymmetric direction of time—*arrow of time*—as well as with the second law of thermodynamics which holds that nature tends toward increasing entropy. Although widely accepted in physics, these important principles have not yet been accounted for and integrated fully into classical or quantum theories.

To summarize, initially quantum theory brought the conscious observer into the picture in the model of wave function collapse. Though the many-worlds and objective reduction interpretations took it back out, it was brought back in as a necessary part of a logically consistent empirical science. And the consistency of experience must not be only within the observer but also between observers for *scientific consensus* in a rational, empirical science.

These points suggest that the classical objective approach that didn't include the subjective observer was inadequate all along. *Objectified* classical science tacitly assumed and freely applied logical consistency within and across observers as core to the scientific method, without acknowledging its implications with respect to mind.

However, we don't yet have a model of conscious mind as ontologically real, necessary for it to interact with the world. The next interpretation includes this major point in a bigger picture of nature that explicitly goes beyond the physical. But before exploring beyond the physical, the key finding of *nonlocality* needs to be discussed, because in this next interpretation conscious mind is real but also *nonlocal*. With mind as real, it could have real power to cause change in nature—not just be epiphenomenal or non-existent and thus powerless. But as real, it needs to exist somewhere and be something.

From locality to nonlocality

After years of debate, experiments were designed to test whether there is an indeterminate random component fundamental to nature, or there are 'hidden variables' that account for the indeterminacy as argued by Einstein and colleagues. [11] Actual experiments were conducted in the 1980s based on *Bell's theorem*, [12, 13, 14, 15, 16] which included assumptions that nature is deterministic, exists objectively independent of the observer, and light-speed sets an absolute speed limit even for any form of information. The results validated *quantum entanglement*, the phenomenon of highly correlated behavior of particles after they interact and separate, though the limitations of light-speed would have disallowed any causal effect on each other.

Nonlocal quantum reality. It turns out, however, that the results didn't show whether nature is determinate or indeterminate. The results were interpreted as showing that the view of objects as interacting only within light-speed is inaccurate. Einstein's view was that all motion is an unbroken determinate chain of physical events within the *local* casual nexus limited by light-speed. These tests showed that *nonlocal* interconnections must be a fundamental feature of the universe, in addition to local features on its ordinary surface. [10]

The finding of the *nonlocal* fabric of nature also could be interpreted to mean that there is a background to the relativistic spacetime gravitational field. But the background has different defining features compared to the model of local physical causality. Objects are not just localized, as long believed in both Newton's and Einstein's classical physical theories. The change from only locality to nonlocality as an additional fundamental feature of nature is another huge step toward *quantum reality* beyond physical reality.

Neorealist interpretation

The final interpretation discussed here integrates nature much more completely. It posits a real nonlocal mind with real causal power over matter. It is consistent with the 'hidden variables' notion of Einstein, but not within the constraints of physical theory. Associated with mathematician and physicist David Bohm, [9] it has been called *neorealism,* due in part to its renewed emphasis on determinism rather than quantum indeterminism. Importantly, the classical scientific tenets of objectivity as independent of the observer and of nature as determinate rather than fundamentally random—initially lost in quantum theory—are both recovered in neorealism. It is a mathematical theory of the motion of particles in which the path of a real matter particle is guided by a real nonlocal wave.

In other words, orthodox quantum theory is incomplete, as Einstein asserted. But although intuiting an abstract field as the basis of matter, he didn't explicitly take the step beyond the physical to a deeper ontology. In the last months of Einstein's life he talked extensively with Bohm, however, which might have contributed to this expanded ontological theory. Bohm also drew from the completely holistic ancient Vedic model in developing this theory, discussed at length in upcoming chapters. [17, 18]

Though the mathematical details are not all worked out, it has been proposed as the most successful interpretation of quantum theory because it provides a solution to the measurement problem and related paradoxes of instantaneous wave function collapse. [19] But it does this via the radical addition of an ontologically real level of nature that underlies, permeates, and causally guides physical matter.

In this interpretation a single quantum object is unpredictable due to classical uncertainty—meaning lack of information—not to an 'inviolable wall' of quantum indeterminism and randomness. Also, the process of measuring by a conscious observer does not create quantum wave function collapse into a classical physical object.

Electrons for example are particles whether measured or not. Their dynamic attributes of motion are guided in part by a real but nonlocal guiding wave or *psi wave* in which they are embedded. To match the behavior of objects in quantum probability predictions, the psi wave must be connected to every particle in the universe, classically invisible, faster than light-speed, and common—that is, a more interdependent, entangled, underlying field or level of nature. [7]

The particle does not explicitly have a wave nature, but is embedded in the subtler psi wave that produces wave patterns along with the known fundamental forces. Together these changing influences produce jittery motion. Because the psi wave doesn't drop off with distance in the manner of physical forces and is highly interconnected (nonlocal), experimental arrangements are so complex as to be *unfathomable*. Thus the particle's path is not precisely predictable—a deterministic account of quantum indeterminism.

The observer's mind is real and nonlocal. The experimental verification of nonlocality logically could not within physical theory account for the mechanism of phase entanglement. The neorealist interpretation provides such a mechanism in the psi wave. The psi wave carries "active information" [17] in that it sensitively reflects the experimental arrangement. A level of nature is posited that is both more extended and more interconnected than relativistic spacetime.

The psi wave is associated with a subtle, real, non-physical, nonlocal field that constantly enfolds and unfolds. Nature is viewed as a single wholeness enfolded in each individual region of space— somewhat like holographic concepts discussed in the next chapter.

Conscious mind is brought back into the picture in terms of the psi wave, influenced by subtle intentional action of *nonlocal mind*. As strange as it may seem, this appears to be the first time in modern science that there is a logically consistent model with means for your real physical arm to be guided by your real mind.

In this model nonlocal mind is subtler than the local brain, associated with the psi wave. But this doesn't mean it cannot be *individual* mind; it doesn't mean something like group mind without individuality. The subtler wave field including nonlocal mind is more specific and more interdependent *at the same time*. It is not discrete like particles, but a subtler and more integrated texture or fabric of nature underlying the relativistic spacetime gravitational field.

The concepts of space and time don't break down beyond the physical, but their features are different. Levels of nature are defined by the limiting features of the medium or substance of which they are built. There is a grosser substance associated with ordinary relativistic spacetime, and now a theorized subtler spacetime substance with fewer limitations—sort of a matterstuff and a subtler energy of mindstuff, but a more abstract nonlocal field notion of energy than in Einstein's equivalence equation which didn't include nonlocality. In this new view the subtle level includes a thought field that permeates the brain/body, and somehow connects with the neural network to guide physical behavior via thought waves of mental intentions. In Part II we will examine aspects of the Vedic model with much more detail about how this non-physical field links to physical matter.

Nonlocal causal wave

In the neorealist interpretation, the classical physical level is the *explicate order*. It is the familiar world of mechanical particle interactions of the four fundamental forces of nature within local causality and the particle interaction model as well as the grosser ordinary 'billiard ball' causal model. The subtle level, the *implicate order*, is a highly interconnected level of nonlocal wave-field interactions that includes mind and abstract thought waves rather than just localized matter and force particles. [18]

Again using the simple iceberg analogy, this time the entire physical universe can be likened to an iceberg in the unbounded ocean. The iceberg is a more restricted state with its own emergent properties, but is made of water. The waves and currents in the water move the iceberg around in the ocean. The environment inside the iceberg is different from the outside water, but basically is permeated by and is the water. The iceberg world is a less dynamic level than the ocean of water from which and within which it emerges.

On the subtle level of nature, causality can be viewed as more wave-like, not particle interaction-like. It is a more abstract fabric of spacetime than the relativistic spacetime gravitational field, with more abstract causal dynamics. But importantly, it *permeates* the gross level and can directly influence it, from deeper inside so to speak.

Although we ordinarily sense events in the physical as discrete in space and in time, we also have an intuitive sense of the unseen converging precedents that causally shape the ordinary events. And as well we have some sense of the continuing aftereffects of the perceived discrete event that shape subsequent processes and events, a richer interdependence underlying and shaping the surface discrete sensory events. All these influences can be envisioned in terms of a *causal wave* that is more spread out in an underlying medium of spacetime, with the gross field embedded in and emerging from this more encompassing and extended subtle field. This would mean there is information in the subtle nonlocal causal wave shaping upcoming discrete events that precedes ordinary information we associate with discrete events in the physical causal nexus.

The discrete, fragmented physical model of independent objects is a more limited sub-set of the subtle interdependent or entangled level that permeates it. The sensory attributions we give events in gross physical spacetime match or correspond to the limitations of that level of nature. The perceptions match the texture of the field from which they are constructed. It is not just that objects and events in space and time change at the subtle level of nature. Rather, space and time, objects and events, and phenomenal experiences of them for the most part correspond with each other in their respective gross or subtle fabrics of nature. The subtle level can influence the more tangible gross level via a more interdependent nonlocal causal wave. As physical objects become more coherent and wave-like, more wave-like phenomena appear, akin to superfluidity or superconductivity.

This causal wave model may be a bit more tangibly pictured in the example of research on the phenomenon of presentiment. This phenomenon relates to anticipatory responses to an event prior to experimental decisions that select the event, presumably based on subtle information not apparent in the physical causal chain, but part of a subtler nonlocal causal wave. It can be likened to the spreading out of spacetime dilation in relativity theory as light-speed is approached. Entangled processes in the subtler level frequently are not apparent in the discrete causal events of our ordinary phenomenal experiences, but can provide subtle information. [20]

Local and nonlocal spacetime. The key finding of the relativistic nonlocal fabric of nature contributes to the important understanding of spacetime in terms of different levels of existence. The gross level is our familiar space and time that appears to have independent objects in it. The much less familiar interdependent subtle level permeates the gross level and is its ontological basis.

In other words, there are concentric fields that are increasingly abstract and subtle, entirely encompassing, permeating, and comprising grosser levels—similar to how space permeates physical objects on the ordinary gross sensory level of nature, only subtler. As might be expected for domains of nature with different fundamental properties, the causal dynamics correspondingly also are different.

Although physical space was thought to be empty, now it can be understood as containing and composing everything in the physical universe. This was a key recognition coming from general relativity theory, described earlier. In turn the subtle level of nature encompasses and is the basis of the gross level. This key recognition can be viewed as coming from quantum theory. And again in turn these levels are based in the infinite eternal unity of nature associated with what Bohm called the 'universal plenum,' also sometimes called the *super-implicate order*. This can be attributed to be a key recognition of unified field theory.

As we will consider further in upcoming chapters, but perhaps difficult to envision, the unified field perspective stretches the picture of spacetime to infinite space and eternal time. It extends to a transcendent field that integrates point and infinity (space) with instantaneity and eternity (time). In the unified field perspective, space and time in terms of the already existing infinite eternal expanse of the unified field can be said to be the stage upon which the play of nature phenomenally appears. It is not that no sense of space and time exists prior to the physical. Rather, more tangible levels of the infinite eternal unified field of spacetime phenomenally emerge in it, from the subtlest to the grossest levels.

The neorealist interpretation of quantum theory is quite different from both the other quantum theory interpretations and the physical theories. Its validity would not mean that these other views are wrong, but rather that their range of external validity is more limited. They can be said to represent relative contextual knowledge of nature at different ontologically real levels or fields of existence.

The neorealist interpretation can be viewed as the beginning outline in modern science for a logically consistent model of how matter is causally linked to our minds. In simple terms it adds subtle spacetime including *nonlocal mind* from which all objects within it are built, expanding on major steps of integration that can be viewed as essential contributions of relativity and quantum theories.

Again, in current mainstream physical theory there is no place where a real mind exists with mental intentions that could cause real change in nature. And consistent with this view, quantum theory further assumes that conscious observers somehow evolved from random, inherently fluctuating quantum fields as the basis of the physical. But extensive research has not found a real place for mind in either physicalist or orthodox quantum models. Fortunately, however, cutting edge progress we will soon overview can be seen to be in the direction of *real quantum mind* beyond the physical, consistent with the advances in neorealism.

As overviewed so far, theories on the forefront of modern science have taken huge steps toward a more expanded and integrated view of nature. In even more integrated and ontologically expanded emerging views, the subtle level underlying and permeating the gross ordinary physical level is determinate (not indeterminate and random); nonlocal (not just local); wave-like (not just particle-like); includes mind as real (not epiphenomenal); includes mind as subtler than matter (not emerging as a product of the brain); includes mind as potentially causally efficacious (we are not zombies or robots with no free will); and includes nature as both subjective and objective (not accepting subjectivity as just separate from and independent of objectivity, or epiphenomenal and not really existing at all).

Cutting edge quantum physics, as well as neuroscience and consciousness studies, is now focusing on how to picture this subtler nonlocal field, and in some cases how to relate it to conscious mind and the local brain. This also is relevant to developing theories of the unified field of everything to be discussed soon, which necessarily would include both objective and subjective.

But first we will consider recent efforts to link relativity theory and quantum theory in *quantum gravity* toward unified field theory. Later we will consider how these theories are converging on the completely holistic model of nature in ancient Vedic literature, which contains three ontological levels of nature with considerable detail about the links between the three levels: unified field (universal plenum or super-implicate order), subtle level including mind (implicate order), and gross matter (explicate order).

Chapter 3

Quantum Gravity

Quantum gravity theories are a key step toward unifying quantum theory and relativity theory into a unified field theory. *String theory* is widely believed in the mainstream physics community to be the best direction for a viable theory of quantum gravity. One mathematical string vibration pattern matches the graviton particle, which may link gravity and quantum theory. However, string theory tends to follow the tradition in classical physics of not addressing the mind. *Loop quantum gravity theory* can be viewed as a more explicit attempt to bring mind into the picture. These theories will be examined a bit more here, because it is in this area that major attention is now being placed and the necessity of addressing how subjective mind relates to objective nature is becoming clearer.

String and M-theories, higher dimensions, and hidden sector

As mentioned in Chapter 2, the concept of a point particle used in physical theory is a dimensionless point with no internal structure, and only the capability of motion through space. But in the mathematics of ultramicroscopic scales in quantum theory, infinite values of energy come into the equations as the dimensionless point is approached. This is inconsistent with respect to calculations of real physical objects, but with no reasonable way to eliminate them.

What is a string? A core innovation in the concept of strings (similar to ideas in ancient Vedic literature) is to replace the dimensionless point-particle with a one-dimensional filament or string about the Planck size. A string is conceptualized as having spatial extension, which allows higher-order patterns of fluctuation that add explanatory power. It also eliminates the problem of infinite values of energy. The higher-order string fluctuations are significant at ultramicroscopic scales, but can be treated mathematically as dimensionless points at coarser-grained scales in classical physical theories. But the change to strings dramatically increases the complexity of the mathematics.

Higher dimensions. String theory generally requires mathematical *dimensions* in addition to the ordinary four dimensions of space and time. Sometimes they are conceptualized as hidden spatial dimensions curled up or *compactified* in the internal structure of the string. The classical four dimensions are described as the non-compactified or *unfurled* dimensions of ordinary space and time.

The notion of a string with spatial extension initially related to the classical notion of space and time, in the sense that there needs to be space and time in which strings vibrate. The mathematical requirements for additional spatial dimensions can be said to imply notions beyond classical space and time. But integrating string theory with relativistic spacetime theory has been quite difficult, suggestive that there may be more to the picture. Again, however, string theory is a mathematical model, and the relationship between strings as mathematical concepts compared to ontologically real objects is not yet articulated, though strings might sound like real physical objects.

String theory is so complicated that its exact mathematical equations have not yet been able to be determined. Approximations yield many models, but there are indicators of a smaller set of consistent ones. Recent advances pull them together into the mathematically encompassing framework of *M-theory*. This theory integrates most all of the theoretical progress in physics in the past century, a remarkable and laudable achievement.

What are branes? M-theory most typically includes 11 dimensions: the ordinary four, plus seven compactified dimensions in mathematical space. In addition to one-dimensional strings, the theory posits two, three, and higher-dimensional 'geometric objects' called *branes*—related to membranes. But higher-order dimensions and strings and branes are in imaginary mathematical space, not the same as the four real dimensions of ordinary space and time. [1]

It is of course fine to conceptualize imaginary mathematical objects for purposes of theory construction. But it is quite different when mathematical models are implicitly assumed to generalize to real objects. Again, how to bridge from imaginary mathematics to ontologically real physics needs to be addressed explicitly for a consistent theory of nature that is practically relevant.

The notion of the texture or fabric of space as made up of incredibly tiny strings or branes fits the model of space as quantized, though extending to infinity in an unbounded wave potential. In quantum field theory the smallest possible size of space, and thus the smallest possible width of a string or brane, is the Planck unit. These mathematical objects, when applied to physics, would be membranes that are no smaller than the Planck unit in width, with apparently no notion of at least ordinary space in between them. As physicist and unified field theorist John Hagelin (p. 10.2) [3] states:

"Once time and distance become ill-defined, the classical notion of causality has no meaning. The three fundamental pillars of rational, deductive logic—time, distance, and causation—have crumbled. Our purely intellectual objective approach to gaining knowledge, which has led us to ever more fundamental space-time scales, has transcended its own methodology—physics has "walked the Planck!".... The Planck scale represents the ultimate scale of physics—there is no currently viable concept of smaller distance scales. It therefore effectively represents the *point value* of physics—the level of *infinite* frequency, and *infinite* dynamism."

Are strings and branes real? Again, quantum models that view space as discrete or digitized in Planck units have been difficult to integrate with relativistic spacetime theory, which views space as analog or continuous. If there is no space between 'geometric objects' such as strings and branes, then how could they possibly be identified as separate, except in some imaginary mathematical way?

As with any real object in nature, strings and branes need a medium within which they vibrate, and in which they can be distinguished. Either they exist in the ordinary relativistic spacetime gravitational field, or they exist in some subtler background medium, or in some way exist in both as sort of a bridge connecting them, or exist as potentia that are not real, or they don't really exist in nature at all other than as imaginary mathematical concepts. The concepts and language to describe quantum objects such as strings seem to go back and forth between imaginary mathematics and ontologically real physics in an unarticulated manner. However, considering the ontological status of theorized concepts again would seem to be a necessary part of the discipline of physics to develop logically coherent models of what we ordinarily call the real world.

In this regard, string and M-theories imply causal interactions between objects in physical space and theorized geometric 'objects' that had been viewed as only in imaginary mathematical space. And the abstract 'geometric objects' are said to form the basis of all physical objects. If so, then they must be real objects at some level of nature. This implies subtler concepts of space, narrowing the difference between ordinary physical space and abstract imaginary mathematical space, even into 'pre-geometric space' outside the relativistic spacetime gravitational field toward mental space, as in neorealism. (Of course, if imaginary *ideas* are real, they also need to exist somewhere, even if beyond ordinary notions of space and time.)

The notion in string theory of closed and open string loops can be viewed as another step toward bridging from local to nonlocal, physical to non-physical. Closed string loops could be thought of as in the direction of the physical, and open loops in the direction of the non-physical. Strings and branes then might be conceptualized in terms of a type of boundary between the physical and the non-physical, a conception used in upcoming models.

A new model to be discussed toward the end of this chapter views relativistic spacetime and gravity, and also closed string loops, as emergent properties. This means that they are background dependent on more fundamental aspects of nature. This new model applies the holographic principle, first describing holographic screens from which spacetime emerges in terms of the boundary of black hole horizons, and then to particles generally as representing different phenomena in a single causal system. The single system is attempting to be envisioned in terms of a more abstract information field that is the basis of ordinary physical aspects of nature emerging from it.

String theory and M-theory can be viewed as at least a little bit further toward bridging the gap between physical space and mental space. But these theories initially developed within the framework of the classical physical theory of space and time as an independent background to matter, and the concept of higher-dimensional space doesn't yet bridge these two concepts of space. In a theorized 'pre-geometric' space, somehow the notion of the Planck scale in terms of ordinary space is maintained. This seems to assume aspects of the ordinary geometry from which Planck unit dimensions have been derived, rather than 'pre-geometric' space beyond ordinary notions of dimensionality. Nonetheless, the theories are clearly bordering on conceptualizing nature in terms of more abstract processes that underlie and generate our concrete familiar physical world.

There is some debate, however, whether string theory and M-theory represent the best direction for a theory of quantum gravity, as well as whether super-symmetry exists in the finite universe in terms of mirror particles/sparticles upon which these theories as now conceived are based. Theories are positing a potential additional background to strings and branes suggestive that they are emergent phenomena and may not constitute a fundamental level of nature.

Zero branes beyond ordinary space and time? A recent advance in M-theory can be viewed as a bigger step beyond ordinary physical space and time. It posits a non-commutative geometry that implies a field of *nonconventional* space underlying strings and branes, possibly existing prior to the big bang to be discussed in an upcoming chapter.

In the following quotes on this advancing research, physicist and string theorist Brian Greene [21] hints at an ontologically real level of nature 'below' the Planck scale and beyond ordinary, conventional notions of space and time, but also *not yet* the ultimate unified field. It is associated with the concept of 'zero-branes' that appear to relate to an ontologically real pre-geometric space that does not seem to fit within classical notions of dimensionality and ordinary spacetime:

> [W]hereas strings show us that conventional notions of space and time cease to have relevance below the Planck scale, the zero-branes give essentially the same conclusion but also provide a tiny window on the new unconventional framework that takes over. Studies with these zero-branes indicate that ordinary geometry is replaced by something known as non-commutative geometry.... In this geometrical framework, the conventional notions of space and of distance between points melt away, leaving us in a vastly different conceptual landscape.... [I]t gives us a hint of what the

more complete framework for incorporating space and time may involve.... Already, through studies in M-theory, we have seen glimpses of a strange new domain of the universe lurking beneath the Planck length...." (pp. 379-387)

"If string theory is correct, the usual concepts of space and time, the framework within which all of our daily experiences take place, simply don't apply on scales finer than the Planck scale.... As for what concepts take over, there is as yet no consensus. One possibility...is that the fabric of space on the Planck scale resembles a lattice or grid, with the 'space' between the grid lines being outside the bounds of physical reality... Another possibility is that space and time do not abruptly cease to have meaning on extremely small scales, but instead morph into other, more fundamental concepts. Shrinking smaller than the Planck scale would be off limits not because you run into a fundamental grid, but because the concepts of space and time segue into notions for which "shrinking smaller is...meaningless... [A]lthough you can divide regions of space and durations of time in half and half again on everyday scales, as you pass the Planck scale they undergo a transformation that renders such division meaningless... Many string theorists, including me, strongly suspect that something along these lines actually happens, but to go further we need to figure out the more fundamental concepts into which space and time transform." *(pp. 350-351)*

Hidden sector? Toward levels of nature beyond the ordinary physical level, one of the models in string theory from which the ordinary 'visible' universe is said to be derived suggests the possibility of a *hidden sector*. This refers to a potential field or group of fields that would be hidden in the sense that they interact only with the gravitational force and not with the other forces that comprise 'visible' objects in the physical universe.

Mathematical calculations further indicate that the theorized hidden sector may have weak interactions with the electromagnetic field. [3] This suggests that it could be at least somewhat visible under certain conditions—that is, not entirely hidden to ordinary experience. It has been speculated that the hidden sector might relate to a subtler but partially visible sector of the universe where mental space and other subtle phenomena could exist. [3] This is in the direction toward subtle levels where the mind might exist in an additional ontologically real level of nature.

On the other hand, the hidden sector would not seem to be a *subtler* level if it is subject to the limitations of the gross relativistic spacetime gravitational field. This reflects a potentially important distinction between interactions with the relativistic spacetime gravitational field versus existing in and fully subject to this field's limiting features, which include light-speed, Planck unit quantization as well as other core principles such as thermodynamics.

Pre-quantum space underneath the Planck scale? The neorealist interpretation of quantum theory, however, explicitly posits a subtler ontologically real nonlocal field—the implicate order. Consistent with this interpretation, in this next quote Bohm [17] points to a real field underneath the Planck scale and the four fundamental force fields. This field would be a background to conventional spacetime that can interact with it but would not be subject to its limitations:

> "[T]he current attempt to understand our 'universe' as if it were self-existent and independent of the sea of cosmic energy can work at best in some limited way... Moreover, it must be remembered that even this vast sea of cosmic energy takes into account only what happens on a scale larger than the critical length of 10-33 cm [the Planck length]... But this length is only a certain kind of limit on the applicability of ordinary notions of space and time. To suppose that there is nothing beyond this limit at all would indeed be quite arbitrary. Rather, it is very possible that beyond it lies a further domain, or set of domains, of the nature of which we have as yet little or no idea." (p. 244)

These glimpses of a potentially real background to conventional space—possibly a "pre-quantum' space" [22] more fundamental than the Planck scale, the relativistic spacetime gravitational field, and the big bang—reflect yet additional steps toward an expanded ontology of nature underneath the physical. The model of different levels of nature with different textural qualities of spacetime can accommodate the need for some kind of substrate that is quantized, if quantization is thought of as an emergent quality that is a limitation of an underlying continuous nonlocal field not defined specifically in terms of Planck unit quantization—but also not yet the supersymmetric unified field.

In other words, the *quantum principle* can be understood more abstractly to have meaning with respect to subtle distinctions within an ontologically real field underneath the physical. But these subtle distinctions would not be in terms of Planck units and the corresponding metrics of conventional spacetime. The next theories discussed are clearly moving toward a more abstract conception of a real information field, but they still apply the quantum principle in terms of Planck scale quantization.

Loop quantum gravity theory

While neorealist quantum theory explicitly places the real mind of a conscious observer into an underlying nonlocal, nonconventional field of spacetime, string theory and M-theory don't seem yet to have addressed the issue directly in pursuing a consistent model of quantum gravity. Another related theory attempts at least a little more explicitly to address the role of the observer in quantum gravity. This approach, *loop quantum gravity theory*, posits abstract non-physical quantized information space permeating and generating conventional relativistic physical spacetime and quantized space. This is a real, finite relative field, but also not yet the infinite unified field.

In attempting to address the observer in terms of observer dependence as in relativity theory, it brings out more clearly the limitations of physical and quantum theories as currently conceived. Observer-dependent relativity theory is trying to be integrated into quantum information space which sounds more mind-like, but as we will discuss it does not yet incorporate fully an actual observer.

Initially an alternative to string theory and now in part apparently being integrated into it, loop quantum gravity theory posits a *pure geometry of quantized information space* more fundamental than conventional spacetime. It is associated with the concept of abstract information bits or *qubits*. This quantized information space is held to generate ordinary spacetime, which means that they must interact as real. Conventional space is said to emerge from topological relationships in a dynamically evolving network of intersecting loops, the *spin network*. [23] The spin network generates curved relativistic spacetime and local particles, sometimes called spin foam, similar to spacetime foam. It can be viewed as taking things a tiny bit further toward an underlying abstract, nonlocal, non-physical *pure geometry* that generates conventional spacetime.

Information space. Adding principles from black hole thermodynamics, the spin network is linked to the concept of bits of a quantized pure geometry of information or qubits in a formal mathematical relationship, *Bekenstein's bound.* [23] Accordingly, the smallest possible surface area of space has an inherent mathematical limit to the amount of information it can contain. Bits of information are directly proportional to an area of ordinary space in Planck units: encoding one bit of information requires one square area of Planck units.

This could be viewed as describing an ontologically real non-physical information space, and thus further toward bridging the gap between the physical and the non-physical by associating quantum bits of physical space with non-physical information bits. [24] But the correspondence of space bits and information bits still doesn't make it beyond conventional relativistic spacetime. It would go further beyond conventional relativistic spacetime if information space is not thought of as being subject to principles of ordinary spacetime such as thermodynamic principles, but it doesn't directly consider this key issue.

In this theory the concept of information remains *objectified.* [25] It draws from information theory which associates information with digital bits of data. This definition doesn't include the subjective sense of semantic *meaning* associated with subjective mind and our usual sense of meaningful information in human intelligence. In the model information is like matter or energy, a functional structure with certain attributes but not inherent meaning or intention. Matter and energy are objective, whereas meaning and intention are typically considered subjective and not included in this objectified meaning.

In usual terms the concept of information connotes semantic meaning, intentions, recognition of significance, and answers to questions about real causal relationships. Associated with the causal system of interacting light cones, loop quantum gravity theory also involves the concepts of causal relations and observer perspectives, and thus can be seen as associated with systematic selective transfer of impulses of energy and information from one to another part of the causal system. In this framework the concepts of intention and meaning seem quite relevant, but are not addressed even in attempting to incorporate the conscious observer into the closed causal world system.

Observer dependent background. Interestingly the relativistic concept of the light cone can be said to provide a clear basis for locating an actual observer in relation to the causal system. The light cone specifies the range of possible influences in an observer's causal world. By definition, the observer is located at the beginning of his or her future light cone, as well as at the end of the past light cone. Thus a real observer exists *in between* these personal past and future light cones. If in typical reductive fashion smaller and smaller scales of the transition between past and future light cones of an observer are examined—getting down to the essential core—what is the essence of the observer perspective and frame of reference, presumably in which can be located the actual observer?

The transition between the past and future light cones can be thought of as a boundary, or a kind of screen. In this view, however, the boundary or screen involves a real observer with a perspective. Reductively analyzing the nature of the gap or boundary formed by the perspective of an observer, in loop quantum gravity as well as a new model to be discussed next, the observer ends up slipping into somewhere else outside the boundary between past and future light cones. This is suggestive of an observer dependent background not accounted for in these theories of spacetime and the classical view of causal relations.

Relativity theory is said to be observer dependent in the sense that the outcome of measurements of spacetime depend on the perspective of the observer. This is associated with the concept of

Lorentz Invariance, which states that regardless of the speed between two objects—such as observer and observed—the laws of nature that describe them will be the same. It is thought to be meaningless to determine which is moving, because motion has meaning only in relation to each other. It is said to be demonstrated for example by the phenomenon that two observers who are close to each other cannot distinguish between the effects of gravity and the effects of accelerated motion—the *principle of equivalence*, placing the reference frame central to the relational nature of spacetime.

To have an *actual* frame of reference that an observer experiences, a real observer is needed—as well as some way to observe, a process of observing. In this sense relativity theory can be seen to contain an implicit psychological theory of mind that assumes local limits to human sensory perception. This issue will be discussed in later chapters with respect to the ordinary waking state of consciousness and corresponding experience typical of the physicalist worldview.

The concept of the event horizon in black hole thermodynamics also relates to the perspective of an actual observer. Not examined is what constitutes an observer in the concept of a *frame of reference* or *observer perspective*. This is comparable to the issue of what constitutes a measurement or observation in quantum theory.

The functional ability to act on information—as in sending and receiving meaningful information and causally affecting events intentionally—seems not to be a part of the objectified observer perspective or frame of reference assumed in loop quantum gravity theory at least with respect to information space. Intentional agents with means to take action don't appear to exist at the smallest time and distance scales where information space enters the picture. If there is no agent at this level that receives, sends, and acts on the information, then the information would not be functionally *informative*—merely abstract information bits devoid of meaning.

The concept of background independence generally means that there is no underlying field or substrate of spacetime that serves as a reference for measuring motion or change. But in positing more fundamental levels than our ordinary four-dimensional physical world, possibly in extra dimensions, contemporary theories imply a background *dependent* substrate to the physical. Observer dependence similarly implies some form of background or substrate. The universe is observer dependent at least in the sense that an observer is needed to observe it and know that it exists; there is at least an experiential background to the universe we observe—namely *us*, the observers. A closed cosmological system needs to account for this kind of observer dependence. Quantum theory places the observer *outside* the probabilistic quantum descriptions of nature. But in relativity theory, the observer must be *inside* the system, while also subjective is still independent of objective—a fundamental quandary.

It is at least quite difficult in these theories to include in the observer perspective the typical features associated with an actual conscious observer. According to the physicalist worldview, an observer's mind and consciousness, usually associated with having a perspective, are products of electrochemical and molecular activity in an individual brain. Since brains exist and function at much larger time and distance scales, the implication is that there are no real observers at ultramicroscopic levels or beyond to information space.

Perhaps mind and consciousness are not necessary in order to have an observer perspective. But then what is it that delineates a particular set of historical events into a perspective or frame of reference? What is the basis for dividing them into individualized partial historical perspectives—such as causal light cones—with continuity through time at a fundamental level of nature where observers with brains and perspectives presumably don't exist?

Limited observer perspective. But loop quantum gravity theory does contain the concept of an observer as something that has a conscious mind. It is also at least recognized—if not directly examined—in orthodox, mind-worlds, and objective reduction interpretations. What would an individualized consistent historical perspective, an observer perspective, mean without observers with conscious minds?

This also brings up what *causes* change in the closed causal nexus. Is the universe a network of constantly changing processes or events that fundamentally are random—with no conscious observer and no agent, either individually or universally—and in which change or information flow is meaningless and devoid of any purpose or value?

If the concept of the observer has any relationship to ordinary meanings such as the ability to act, function intelligently, reflect, be aware of, evolve, and so on—which the concept of the causal light cone seems to imply—then the system couldn't be observer dependent in the sense of relativity theory when applied to loop quantum gravity. This is because presumably there are no observers to depend on with perspectives at the tiniest ultramicroscopic scale of nature where information space is theorized to enter the picture.

In contrast to relativity theory, observer dependence here would seem to contain only the concept of a separate history or channel—the notion of a part or bit of the entire cosmological system—not the observer as an experiencing conscious mind. The closed system in relativistic loop quantum gravity apparently has no agent that either experiences or causes the flow or change of anything at fundamental levels of nature, and information space has no *meaningful* information in it. To have meaningful information and mental intentions with causal power—as in neorealism—the classical notion of an observer as an emergent macroscopic biological system is inadequate.

But again, at least to a limited degree loop quantum gravity theory does contain the concept of an observer with a conscious mind. It brings the mind of a conscious observer into the picture in an attempt to *relativize* quantum theory. It interprets the relativistic frame of reference as a partial consistent history of the universe from a particular observer perspective. *Quantum superposition is taken to be at the level of the mind*, in terms of overlapping observer perspectives.
[43] Superposition of histories, each independent and associated with a different observer perspective, allows for agreed-upon outcomes with the same initial conditions and similar consistent histories.

Though the conscious mind of a real observer is included, it is incomplete. The conscious observer still experiences a unitary state from outside the causal system, as in quantum theory. But in loop quantum gravity theory as well as relativity and quantum theories, there is supposed to be nothing outside the relativistic causal system. Core aspects of the conscious observer—especially self-observation or self-reflection—are left out too. As loop quantum gravity theorist Lee Smolin [25] explains:

"The quantum description is always the description of some part of the universe by an observer who remains outside it.... If you observe a system that includes me, you may see me as a superposition of states. But I do not describe myself in such terms, because in this kind of theory no observer ever describes themselves. Rather than trying to make sense of metaphysical statements about their being many universes—many realities—within one solution to the theory of quantum cosmology, we are constructing a pluralistic version of different mathematical descriptions, each corresponding to what a different observer can see when they look around them. Each is incomplete, because no observer can see the whole universe. Each observer, for example, excludes themselves from the world they describe. But when two observers ask the same questions, they must agree.... One universe, seen by many observers, rather than many universes, seen by one mythical observer outside the universe." (pp. 47-48)

This reflects the continuing challenge to go beyond the classical physicalist view of the conscious observer as a product of the gross local brain. The ontologically real place for an abstract, non-physical field of information space that could generate physical spacetime would need to be more fundamental than conventional spacetime, and possibly *underneath* the Planck scale. But still, the observer's experience remains outside of information space, as well as outside the entire causal world system as it is attempting to be described in this theory. While the observer's perspective is said to be inside the causal system, core features of the observer including actual subjective experiences related to a perspective, and also self-observation, still remain outside—again suggesting a bigger picture.

To summarize this overview of quantum gravity theories from the unique angle of how mind fits into them, the overall picture is of smaller time and distance scales to the Planck scale, hypothesized to be the ultimate limit of space and time as we ordinarily think of them. This scale is so incredibly tiny as to be *almost* a dimensionless point. It is posited that compactified near about this smallest size are geometric 'objects' in hidden compactified dimensions that form discrete particles in our real, four-dimensional visible universe.

These terms may give the impression that the abstract 'geometric objects' are physical and real. But they are mathematical objects described using spatiotemporal metaphors to help develop the theories. They don't yet bridge the gap between concepts in imaginary mathematical space and ontologically real physical objects in the conventional spacetime gravitational field.

The notion of zero-branes takes it a step further beyond conventional notions of space and time into a 'pre-geometric' space not based on classical notions of dimensionality. If abstract objects in string theory and M-theory have ontologically real existence in a 'pre-geometric' space and are not just concepts in mathematical models—like quantum waves in objective reduction interpretations—then they would be fundamental curvatures of space and time that are at least beyond ordinary matter. And if they are generated from abstract information qubits—like in loop quantum gravity—then there would be an underlying quantized information space that is closer to being beyond relativistic spacetime and Planck-unit physical quantization.

Loop quantum gravity theory also importantly begins to address the consistency of the observer's experience and consensus across observers, related to consistent histories and initial conditions. However, although the theory is said to include the observer in the relativistic causal system, the observer's experience still is not accounted for in the system—remaining outside. In addition, the notion of information space is not integrated with the perspective of a real conscious observer that is tacitly assumed in relativity theory.

An ontologically real information space has affinity with terms such as imaginary space, mathematical space, conceptual space, configuration space, field space, phase space, hyperspace, superspace, seed space, higher dimensional space, hidden dimensions, nonconventional space, pre-geometry, and pre-quantum space—in the direction toward mental space and quantum mind. These terms reflect attempts to characterize an ontologically real level of nature more fundamental than the conventional relativistic spacetime gravitational field. The difficulty of integrating relativity and quantum theories into quantum gravity may be because both are conditional and cannot account for this underlying background.

In other words, relativity, quantum, and quantum gravity theories may be background dependent and also observer dependent. They might emerge from more fundamental levels. This also is suggested in the next model, which uses the language of emergence rather than background dependence. With respect to an ontological background, this next model is a bit more abstract and thus toward the Vedic model discussed in Part II. But it doesn't contain any recognition of observer dependence with respect to the more fundamental level.

Importantly the Vedic model views the physical universe as dependent on—or emergent from—a more fundamental background or ontologically real field. It further views these fields as dependent on an even more fundamental and subtler mental field, more similar to the neorealist quantum theory interpretation described earlier.

Relativistic spacetime and gravity emerging from entropy

A recent model by physicist Erik Verlinde [25] brings out interesting views of the relationship between physical space and information space, applying the holographic principle, black hole thermodynamics, and drawing from string theories. This model reflects a developing perspective that Newton's and Einstein's models of gravity do not describe a fundamental force of nature, but rather an emergent 'force' that naturally results from entropic processes. It is a step further toward explaining forces of nature in terms of information, which can be viewed as just a little closer to mind-like properties such as thoughts, intentions, and mental space.

However, the notion of information as used in this model still remains objectified. It is not the type of pragmatic, meaningful information usually associated with mental concepts and subjective experiences; and it does not explicitly include nonlocal processes.

Further, the holographic principle is applied to Planck-unit areas of information space, associated with the Bekenstein bound discussed earlier in loop quantum gravity. This can be viewed as somewhat akin to the idea of a smallest possible physical memory chip. But how the Planck-unit bits of information space become individualized or digitized, as well as features that might relate these bits of information to ordinary meanings of information and an observer perspective, also are not addressed. The information space still does not know anything, including itself, and cannot do anything—qualities that would seem to be important to us, who might have at least some intuitive sense we happen to be real observers of nature.

Another key issue not addressed in the model is that information space, if more abstract than physical spacetime, would seem no longer to be 'hot.' A key notion of Planck scale processes characteristic of matter/energy is that they concern thermodynamic principles and temperature. But the concept of information can be viewed as more abstract and not carrying with it notions of mass/energy/density that would necessarily place it as subject to ordinary thermodynamic and gravitational principles. It thus seems to be a carry-over from relativity and reductive physical theories.

The holographic principle and coarse graining. The holographic principle refers to the encoding of a three-dimensional image onto a two-dimensional space. The objects encoded and the boundary or screen upon which they are encoded are not really two-dimensional, but are thought of in this way for mathematical purposes. Indeed, it can be said that there are no objects with only one or two spatial dimensions in the phenomenally real world. All phenomenally *real* objects are three-dimensional.

Encoding of the information into a hologram is said to be accomplished in part due to coarse graining, in which not all the available information is needed or used in the reductive encoding from three to two dimensions. But the outside surface is encoded, not all the many layers 'inside' the object 'below' the surface. At each layer of a physical object, the three-dimensional surface image at that layer can be encoded on sort of a two-dimensional surface. For example, the form of a human body, as well as the surface of the internal organs and other organs inside the body such as cells, neurons, molecules, and so on could be encoded. Surfaces at smaller scales inside each of these objects or forms theoretically could be encoded too—whatever 'surfaces' might mean there.

But all the layers of an object taken together—macroscopic as well as microscopic and ultramicroscopic layers typically described as deeper inside—have additional information not in the two-dimensional hologram. In terms of layers of the physical aspect of this deeper inner dimension, slices of the layers of the object could be made thinner and thinner, theoretically down to the limit of the Planck scale. More and more of the available information would need to be encoded with these increasingly finer-grained slices of each layer of the object, in order to preserve and represent the information about them in a holographically encoded image.

At the level of the finest graining of the physical, the relationship between the physical quantized space and its mapping onto quantized information space would be the same level of detail of information; that is, there would be no coarse graining in the encoding of all layers of the complete object at the Planck scale level of fine graining. And according to the Bekenstein bound relationship between physical and information space, it seems that there would be no more 'room' to map additional information within the limitations of the Planck area/information qubit correspondence.

But this doesn't necessarily mean there is no more information. The holographic mapping of information typically is static—although methods to encode dynamic processes are being attempted. Considerably more Planck unit bits would be needed to map dynamic change. Where would that information be encoded at the Planck unit level of detail where coarse graining no longer applied?

And much further, if there is a real level of nature that is nonlocal, causally interdependent and entangled, then its information would seem to be far beyond the capability of the Planck area/information qubit correspondence to map. The holographic principle with respect to encoding three-dimensional surfaces on two-dimensional surfaces would then break down. If Planck-width slices of you were encoded, still your unitary self and conscious mind wouldn't be included, according to an expanded ontology of nature. Or if your conscious mind is epiphenomenal, apparently it wouldn't be encoded but might still magically jump over to the encoding?

However, there certainly would seem to be enough room in nonlocal space to map all local information. This suggests that the holographic principle as typically described and implemented corresponds to our physical model of the universe. Although the *holographic principle* might well be valid at real nonlocal levels of nature, it clearly would not have the same limitations with respect to encoding at the gross local level. This suggests that the Planck area/qubit relationship is framed by the limitations of the medium of the ordinary relativistic spacetime gravitational field corresponding to the reductive physicalist worldview. Verlinde [26] asserts:

> "...using only space independent concepts like energy, entropy and temperature, it is shown that Newton's laws appear naturally and practically unavoidably."

In contrast, the suggestion here is that they way in which energy, entropy, temperature, and even information are used is not space independent but rather dependent. They also can be viewed as being derived from the relativistic spacetime gravitational field conceived as a medium or matterstuff, from which physical processes emerge.

It is consistent with this view that coarse graining would be how our ordinary physical senses work as well. It also may relate to the concept of decoherence and the apparent loss of quantum wave interference effects at classical physical scales.

We generally observe the macroscopic three-dimensional surface of whole objects, not all the underlying inner details on finer-grained layers at smaller time and distance scales. Ordinarily we don't see the electrochemical processes involved in the sensory and perceptual analyses of the objects that compose the perceptual gestalt. Only the integrated products of these underlying processes are presented to ordinary conscious attention. Otherwise there would seem to be too much detail for perceptual systems to function in order to guide behavior in the ordinary physical world.

Thus it would not be surprising for a model of nature that emphasizes physical spacetime to yield the Planck area/qubit correspondence. But this doesn't account for a nonlocal level with object interdependence. It is perhaps also useful to note that deeper, smaller time and distance scales of physical objects exist right on the macroscopic surface. Typically they are not seen due to a form of coarse graining, at least in the human perceptual system.

We can take this perspective on the holographic principle to its maximum. An infinite unified field can be associated with an ultimate meaning of the principle, in which each point contains the whole. Every point is infinity, and all information is in each point. This is the unified field as *infinitely self-interacting*. It transcends our intellectual understanding, because it would be beyond all things including our intellects. This is fundamental to the unified field as the source and container of everything in the Vedic approach, discussed in Chapter 8. It is made relevant to our own experience in the following quote from Sri Nisargadatta [27]:

"There is enough space in a point for an infinity of universes. There is no lack of capacity. Self limitation is the only problem. But you cannot run away from yourself. However far you go, you come back to yourself and the need of understanding this point, which is as nothing and yet the source of everything." (p. 337)

The notion of the holographic principle expanded into nonlocal levels also suggests a different perspective on teleportation. Porting by decomposing you into abstract information needs to include the totality of what you are. Even if you are just physical, the information would need to be decomposed and recomposed at all levels of coarse and fine graining, potentially all the way down to the Planck scale.

But even then, only local information would be included. If you are more than physical, the additional information including nonlocal levels where cutting edge models now place your mind would be excluded. Neural correlates of subjective processes have been extensively researched and detailed correspondences are even now being discovered. But mind has not been found in the brain/body, as well as perceptions, memories, unitary self, consciousness, or any subjective experiences, and thus apparently wouldn't be 'ported.'

Also, holography is different from 'porting,' inasmuch as it does not decompose the object but rather just makes an image copy of it. Holography copies information about the object, but does not remove the information from it. It would seem to be the same with respect to theorized encoding of information onto black hole screens. Holographic encoding does not decompose the object.

In other words, we need to 'think outside the box' of the physical for a consistent and integrated view of mind and matter. But at least it is becoming increasingly clear that the physical is not the full picture. The local physical body/brain may be tightly linked to the nonlocal mind as a transducer that receives information waves from the nonlocal mind and transmits them into local physical behavior via neuromuscular processes. Later we will discuss how this might occur.
Black hole thermodynamics, entropy, and information. Black holes were predicted based on general relativity theory when a region of spacetime has enough density of physical matter/energy for gravitational attraction to counter any forces that separate them. When gravity prevails, matter/energy is pulled together and at some point the attraction is so strong that nothing, presumably including any form of information, can escape.

Eventually the increasing pressure increases the temperature and becomes so intense that the black hole may explode, expelling matter, energy, and apparently information outward. These mechanics may be part of the dynamic warp and weave of the relativistic spacetime gravitational field over long periods of time, depending on how matter/energy and possibly information bits get distributed throughout the universe. Black holes at the centers of galaxies suggest different stages of these processes in relativistic spacetime.

Verlinde's model uses the concept of black hole horizons to examine the nature of gravity and its relationship to thermodynamics, entropy, inertia, and information. The horizon is thought of as a boundary between the perspective of an observer outside it and another observer that has fallen into the black hole. The boundary is again conceptualized as a two-dimensional screen. But the holographic principle exemplified using black hole thermodynamics is generalized to anywhere in conventional space and thus concerns fundamental properties of the nature of conventional spacetime itself.

For several years there was a debate by leading theorists as to whether objects that are sort of lost into the black hole, from the outside perspective, would also mean that the information about the objects also would be lost. One side (spearheaded by Hawking and Thorne) asserted that the information would have to be lost, at least in the sense that it can never be recovered by outside observers. The other side (mainly t'Hooft and Susskind) felt that this violated fundamental principles of nature, and thus there must be some way that the information is not lost. Rather, via some unknown mechanics the information is encoded onto the black hole horizon. The debate seems to have been concluded in favor of this latter view, which Verlinde's model employs.

According to Verlinde's model, [26] the force that we have understood to be gravity is accountable for in terms of changes in the amount of information, measured as entropy:

> "The product of the temperature and the change in entropy due to the displacement of matter is shown to be equal to the work done by the gravitational force.... An entropic force is an effective macroscopic force that originates in a system with many degrees of freedom by the statistical tendency to increase its entropy.... The statistical tendency to return to a maximal entropy state translates into a macroscopic force." (p. 3)

In this abstract model, the dynamics of a black hole horizon is used to exemplify how relativistic spacetime as well as acceleration, inertia, and the gravitational 'force' emerge at a macroscopic level with coarse graining. A finite amount of entropy is associated with each configuration of matter, and this entropy is a measure of the amount of microscopic information that is not visible to the macroscopic observer. A direction in space due to coarse graining is shown mathematically to be an emergent spacetime geometry that also accounts for gravity, acceleration, and inertia.

Emergent spacetime and gravity are described in terms of black hole horizons with respect to the perspective of observers outside the black hole. But of course there is another perspective, namely the one from the other observer unfortunately being pulled into the black hole. According to the model, the information about this observer is encoded onto the two-dimensional surface of the event horizon as a holographic screen, and thus it is not lost forever to us on the outside. From the standpoint of the observer going into the black hole, however, apparently no major change would be initially evident.

The information about this observer remains in the observer, though a copy image of the surface of the observer may become stuck to the screen of the event horizon via some unknown mechanism. Presumably this observer eventually will be pulled apart by the extreme gravitational forces as she or he inevitably descends toward the center of the black hole. Apparently information associated with this observer would dissipate, with a copy encoded on the event horizon.

But what about information associated with experiences the observer had after passing through the event horizon? Perhaps this information is somehow encoded in the bits of energy/information that get expelled if the black hole explodes. Alternatively, in some way the information may remain at the black hole core. If so, then this would suggest that the black hole couldn't collapse to singularity, because there needs to be enough qubits to hold the information so it is not lost, and these qubits correspond to Planck-size units of the physical spacetime field.

Back to the model, Verlinde [26] states that:

> If gravity is emergent, so is spacetime geometry. Einstein tied these two concepts together, and both have to be given up if we want to understand one or the other at a more fundamental level....Instead of only focusing on the equations that govern the gravitational field, we uncovered what is the origin of force and inertia in a context in which space is emerging. It is driven by differences in entropy, in whatever way defined, and a consequence of the statistical averaged random dynamics at the microscopic level. The reason why gravity has to keep track of energies as well as entropy differences is now clear. It has to, because this is what causes motion! (p. 22)

This quote seems to suggest that physical motion is 'caused' by microscopic 'random dynamics.' But Verlinde [26] also further acknowledges that:

> "... we are entering an unknown territory in which space does not exist to begin with." (p. 23)

It might be said, however, that this statement refers to our ordinary meaning of space. In the following statement, he notes that:

"...the screens that store the information are like stretched horizons. On one side there is space, on the other side nothing yet." (p. 6)

This statement clearly is suggestive of something more (the 'other side'), which is 'nothing yet' in the sense that we are in unknown territory beyond our ordinary understanding of space. Verlinde's model attempts to show that Newtonian and relativistic notions of space are emergent, and the model acknowledges that it emerges from something more fundamental, in 'unknown territory.'

As to the nature of the unknown territory, the holographic encoding of information from which space is said to emerge requires basically three ingredients in the model: a tendency in nature to maximize entropy, associated with amounts of information; temperature or a thermal environment; and a "bookkeeping device that keeps track of the amount of information for a given energy distribution" [26] (p. 11). Importantly these ingredients can be associated with three fundamental principles in the Vedic model, discussed later. They can be associated with a tendency to resist change (mass) or maximize entropy toward a state of non-change; an activation (energy) tendency, which can be associated with inherent dynamism or a thermal environment; and a tendency to retain order, a negentropic principle of the maintenance of order, which can be associated with a bookkeeping device (information) in the model.

Relationships described by Newton's and Einstein's models of space can be said to emerge via these three ingredients. In other words, it is the quantitative properties of physical space and gravity that are dependent on these three ingredients. But the three ingredients also emerge from some more fundamental level, such as information space or a deeper level of the three ingredients.

In the Vedic model, at the gross level associated with ordinary matterstuff (explicate order), the tendency to maximize entropy can be said to be at its strongest compared to the other two principles or ingredients. The Vedic model describes the deeper origin of the three ingredients or properties with a much subtler interpretation of the ingredients with respect to mental space. A deeper view of these principles will be discussed at length in Part II. How subjectivity and mind of a real observer associated with the subtle level or implicate order link to the gross level or explicate order is explicitly addressed in terms of these three fundamental principles.

To continue our overview, the remainder of this and the following two chapters will examine other theories toward ontological levels beyond ordinary space and time. These theories are attempting to envision a bigger picture in the context of cosmology and unified field theory, further in the direction of the holistic Vedic view of nature.

Dark matter and dark energy

The major focus in recent decades on quantum gravity is due in part to the discovery of mathematical symmetry, applied to smaller time and distance scales in the form of theorized super-symmetric partners of the known particles. In this view the emerging universe broke from its super-symmetric state through *spontaneous sequential symmetry breaking* into the fundamental force fields. This is said to have occurred automatically as the high energy and temperature dropped via the universe expanding outward from the big bang.

One approach toward understanding the unity of nature is to think back in time to events at the 'time' of the theorized big bang—so to speak, attempting to put the cosmic egg of the universe back together. The principle of symmetry has been important for developing theories that describe how—at about 10^{-16} cm—the electromagnetic and weak nuclear forces unify into the *electroweak* force. This level is electroweak unification, and it relates to the *Standard Model*.

There also is evidence that at even higher energy and smaller scales—about 10^{-27} cm—the electroweak and strong nuclear forces unify into the strong-electroweak force, or *Grand Unification*. This is thought of as bringing everything in nature down to two fundamental fields, but not yet a single field—and not yet including nonlocal mind.

The mathematical principle of *super-symmetry* supports theories attempting to unify the remaining two force fields—strong-electroweak and gravity. It requires super-symmetric partners with matching properties to be found for all the known particles, recently associated with the concepts of *dark matter* and *dark energy* based on the mathematics.

Dark matter was proposed in part to construct theories applying the principle of super-symmetry to explain how galaxies hold together, which requires more energy than is available in the 'visible' universe. It is different from *dark energy*, which was proposed to help explain the evidence that the universe is expanding at an increasing rate. But to date, no super-symmetric partners of the known particles have been found that would provide empirical support for string theories, as well as for dark matter and dark energy. Other models are being suggested that don't require dark matter and energy, and that view these concepts as ad hoc hypotheses.

As mentioned in prior chapters, this reflects the tendency in recent years for major theories in physics to be evaluated and even validated on the basis of mathematical consistency more than empirical validation. Experimental tests of the theories are increasingly difficult to implement beyond the ordinary physical world they were designed to investigate.

This is suggestive of an evolving epistemology of research toward increasingly abstract conceptual models of nature and decreasingly tangible empirical means to test them. Mathematical consistency and principles such as symmetry are used as means not only to develop scientific theories but also to test them. This is evident in string theory, the most prominent view in contemporary physics but that also presently lacks empirical support. As Hagelin [3] points out:

> "Superstring theories and M-theory are theories of physics at the Planck scale, and are forever beyond the reach of experimental verification using conventional accelerator technologies. Some have therefore argued that superstring theory isn't physics—that it is just mathematics or philosophy." (p. 115)

Given this trend, some prominent theorists have commented on what might be characterized as a *faith-based* direction in contemporary physics. [28, 29] But of course the hope and intent for empirical validation of the mathematical models certainly remains (of course if subjective processes such as hope and intentions are real).

It also can be viewed as suggesting the need for an expanded epistemology to investigate levels of nature beyond the physical. This will be discussed briefly at the end of the book in terms of a bigger picture of epistemology and the means of gaining reliable knowledge and experience of nature.

Hagelin [3] summarizes progress toward the unification of objectivity and subjectivity that incorporates both reductive and holistic perspectives of nature in an integrated perspective:

"It comes as a surprise to physical scientists that the unified field is consciousness, because consciousness, *per se*, is not in their world view. Three hundred years of scientific investigation into billiard ball mechanics and celestial mechanics has led to an understanding that the universe is inert—lifeless....For these scientists, consciousness must arise as an emergent property of the material brain. But more than a century of Western philosophy, psychology, and neuroscience provide no clue as to how consciousness—how subjectivity—can arise from inert matter.... But there have been revolutions in our understanding of natural law, from quantum mechanics to quantum field theory to unified quantum field theory, which have shown that the material universe is NOT inert—and that it is not even material! The fundamental qualities of consciousness are increasingly expressed at more fundamental space-time scales." (p. 12.2)

The next chapter focuses on the emerging picture of an all-encompassing unified field, and its implications for cosmology. In the final chapter of Part I, we will consider three-level models that are clearly in the direction of the ancient Vedic three-in-one consciousness-mind-matter ontology rich enough to link mind and matter, the theme of Part II.

Chapter 4

Unified Field vs. Big Bang

Reductive big bang theory is sometimes described as suggesting that the universe including space and time began from literally *nothing* (ex nihilo). And then it instantaneously was randomly fluctuating quantum fields of extremely high pressure, density, and temperature that formed the four-dimensional field of space and time and further, through spontaneous sequential symmetry breaking, the other quantum force fields as the basis of physical matter.

Higgs field and mass. To explain symmetry breaking into particles with mass, an additional field, force, or quality has been posited in recent years, associated with the theory of the *Higgs field*. Considered to be one of the most important concepts in the past century in theoretical physics, it is now a major focus in experimental research.

Higgs field theory proposes that in the third phase of symmetry breaking into the weak and electromagnetic forces, a Higgs field condensed to a nonzero value when the temperature of the universe dropped to about 10^{15} degrees, creating a *Higgs ocean*. The Higgs ocean can be thought of as a *viscosity* in space that resists change, giving mass to particles. Thus an inherent quality that resists change is posited in nature, along with inherent dynamism discussed earlier.

In this view it was the blasting out of energy in the big bang and more fundamentally the underlying inherent dynamism of random quantum fluctuations that initiated the causal chain in nature, also producing a viscosity to space leading to objects with mass. Consistent with this view, we as physical objects have no real power to initiate change, because we are just another link in the unbroken causal chain that began long before living beings evolved in nature.

Einstein's theory of general relativity allowed the possibility that spacetime, and thus the entire physical universe, could either shrink or stretch. Because this contrasted with his belief in a static universe, he added another term—the cosmological constant. This allowed the equation to contain a negative value, meaning that gravity could be repulsive rather than just attractive. If carefully chosen, repulsive and attractive forces could balance out, resulting in a static universe. When evidence revealed that the universe is expanding, Einstein withdrew the cosmological constant, reportedly identifying it as his greatest blunder. [30] But it was later revived in the context of Higgs field theory and a modification of standard big bang theory, *inflationary* big bang theory.

Inflationary big bang theory. In this theory, for an extremely brief period of 10^{-35} sec gravity became a repulsive force driving the emerging universe into a colossal expansion. This involved the Higgs inflaton field, contributing a uniform negative pressure to space that produced a repulsive force so strong the universe expanded by an incredible factor as much as 10^{100} more than in the standard theory. [3]

This theorized inflation of space can be thought of as much faster than light-speed, but also not inconsistent with it. The reason is that light-speed applies to motion *through* space, whereas inflation refers to the expansion *of* space. This can be understood to imply speeds faster than light-speed, but not instantaneous, in a kind of space and with a kind of motion that are different from conventional notions— including a *vacuum energy* not powered by ordinary matter and energy. This is associated with the concept of dark energy, sometimes also described as similar to the ancient term 'quintessence.'

Further, it implies a distinction between inflationary space and the gravitational field of relativistic spacetime limited by light-speed. The age of the universe is estimated to be 13.7 billion years, but the estimated radius of the universe is about 48 billion light-years. Along with other implications for a bigger picture than physical theory which have been noted already, these estimates suggest some different kind or texture of space more extended than the relativistic spacetime gravitational field limited by light-speed.

Inflationary big bang theory posits a total amount of matter and energy in the universe that is considerably more than the tally of visible objects, which contribute about 5% of the total. Astronomical research suggested additional matter is needed to hold galaxies together. This led to the theory of dark matter, based on principles of symmetry. It is estimated that dark matter accounts for an additional 23%. Observations that the universe is expanding based on measurements of the recession rates of supernovae helped to revive the cosmological constant, associated with dark energy and super-symmetry. It was estimated that the expansion rate requires a cosmological constant related to an amount of dark energy that contributes about 72% more of the total, which fits the remaining amount in inflationary theory. This 'consensus' cosmological theory also helps explain how stars and galaxies were formed.

But what triggered inflationary expansion? How could literally *nothing* blast out? Indeed, how could nothing do or be anything?

Pre-inflationary big bang theory. An elaboration of inflationary big bang theory proposes a *pre-inflationary* period in which the gravitational and Higgs fields were bumpy and highly disordered, and a random fluctuation resulted in the values needed for inflation. How these fields came to exist, became quantized, became either inherently dynamic, attractive, or resisting of change, and how they could have blasted out with incredibly immense energy and heat are not addressed. But it sure doesn't sound like they came from nothing.

'When' the big bang theoretically began, an orderly temporal sequence also began. At least in the world as we usually understand it through science, an event manifests in an orderly manner from the previous event. This basic scientific principle can be viewed as implying that the source of the universe is a state of order, not *fundamental* randomness, and also a source of at least immense energy if not unlimited energy. This view is crucial for understanding the origin of the laws of nature and the structure of the cosmos. Big bang theory needs to be consistent with unified field theory— reductionism and holism need to match up.

Importantly, in quantum field theory space is *not* empty nothing; it at least contains quantum fluctuations, vacuum energy or inherent dynamism. And in unified field theory, the universe is viewed as created from *something*—even from the *source of everything*.

A key component of super-symmetric unified field theory is that the fundamental force fields emerged through spontaneous sequential symmetry breaking as the universe expanded and temperature dropped. This can be likened to phase transitions of H_2O condensing from steam to water to ice as temperature drops; at each stage, symmetry is reduced. In this view the fundamental forces potentially *pre-existed* in the super-symmetric unified state. And as the source of continuously occurring quantum fluctuations or inherent dynamism, the unified field continues along with the symmetry breaking.

If it continues even after the fundamental forces differentiated, then it is more than only the unification of these forces. The underlying unity—as well as dynamism—doesn't vanish with symmetry breaking. The initial symmetry breaking implies that the unified field itself is super-symmetric. The notions that the unified field is the source of everything, the basis of all the laws of nature, super-symmetric, and the origin of order throughout nature are consistent with the view that it is a field of zero entropy.

Thus in the unified field perspective nature might be said to contain *inherent order* (information). This quality can be added to the other theorized qualities of inherent dynamism (energy) and inherent resistance to change (mass) discussed earlier. They are consistent with the three ingredients or qualities discussed in Chapter 3 with respect to ordinary gravity emerging from entropic processes. But the relationship of these fundamental qualities to the four fundamental force fields seems not yet to be addressed. These three ingredients or qualities as inherent in nature are core to the ancient Vedic account and have been extensively articulated in it. In Part II speculations about how this model of three qualities could be related to the Higgs force and the four fundamental forces will be considered.

Einstein's general theory of relativity suggested that the universe either expands continually, or expands and then shrinks. The estimated total mass energy in the universe adding together all visible objects such as stars, planets, and nebulae fits an expansion model. However, some recent cosmological theories suggest the universe might be starting to shrink, or might have cyclic expansions and collapses over vast eons of time, as described in Vedic literature. [31, 32]

But the big bang could not have started from some point in space and time, if there was no space or time 'before' it in which it could start. And in unified field theory the universe couldn't blast out in a big bang, because everything remains inside it. If the unified field is the lowest or no entropy super-symmetric state, then pre-inflationary big bang theory that holds low entropy came from inflationary expansion would seem to imply the inconsistency that something existed before the unified field.

As the all-encompassing source of everything, no *thing* would exist prior to it, outside it, or create it—not even randomly fluctuating gravity and Higgs fields, or vacuum energy. How the Higgs field emerges, or emerges from the gravitational field, and how the gravitational field emerges, need to be addressed in a logically consistent cosmological model.

Blasting out inside? With respect to the entirety of existence in unified field theory, the big bang would not explode outside from the unified field because everything resulting from it remains inside it. The big bang would not create spacetime, but rather be a limitation of the unified field—perhaps a 'big condensation,' but not a big bang blasting out to create space and time and everything from nothing. In other words, it is helpful to *think outside the bang.*[33]

A similar conception was used in Chapter 1 as a subtler understanding of relativity theory, in which all physical objects emerge from relativistic spacetime gravity. In terms of the unified field, all aspects of nature including subtle non-physical and gross physical emerge within it and are made of it and nothing else.

In this view there can be individual bangs from exploding black holes in the ordinary spacetime gravitational field. The hypothesis that each galaxy has a black hole at its center is consistent with the view that black holes can collapse to near a point, at which the extreme density may result in an explosion outward in a local bang with enough matter/energy. If not, then other astrophysical structures can result. This process of local banging and collapsing might be dynamic pulsations of the relativistic spacetime gravitational field, through which galaxies repeatedly form and collapse.

These long-term cyclic pulsations would be within the fabric or medium of ordinary physical spacetime. But they may not seem to account for evidence consistent with the model of the entire physical universe expanding at an accelerating rate. On the other hand, the processes might be similar with respect to the entire physical universe. An accelerating expansion seems to mean that every point, so to speak, in ordinary space is expanding. Apparently at the smallest possible time and distance scales the expansion would be miniscule, and at the most distant regions the expansion would appear to approach light-speed. This is suggestive of the finite local field as a medium that can be said to curve back upon itself, as characterized in general relativity theory within the limits of light-speed.

But also, models are now being proposed that don't posit dark matter and energy to account for the apparent accelerating expansion of the physical universe. Some even argue for an alternative to the entire model of the universe blasting out in a big bang. [31, 32]

One example emphasizes antimatter as a repulsive gravitational force that may have collected in what are called 'voids,' vast areas in conventional space absent of galaxies or other visible astrophysical objects. [34] According to this model, the matter and antimatter predicted by general relativity theory are repulsive with respect to each other, but self-attractive. Antimatter would collect into large voids with almost no matter in them, and could be powerful enough to drive the accelerating expansion of the universe in possibly a cyclic pattern without dark energy and without a big bang explosion.

Another alternative not requiring dark matter and energy or a big bang is also associated with repulsive gravity. [35] This model states that the accelerating expansion of the universe can be accounted for in terms of the virtual fluctuations of matter and antimatter in the quantum vacuum, producing real repulsive gravitational dipoles in nature of matter and antimatter. The energy density in the theorized gravitational dipoles has been calculated to add the correct magnitude of a cosmological constant for the accelerating expansion. These and other models represent fundamental reconsideration of consensus cosmology that may yield major advances in understanding the totality of nature. They provide alternatives to the consensus big bang theory that at least at this point seem to be plausible alternatives.

Most all the models discussed so far emphasize binary conceptions of nature with fundamental dualities, as in real/virtual, matter/antimatter, particles/sparticles, fermions/bosons, attraction/

repulsion, classical/quantum, conventional space/nonconventional space, actual/potential, unity/diversity and so on that echo the mind/matter duality this book focuses on bridging. Perhaps the starkest of contrasts is the nothing/everything distinction.

Nothing vs. everything. In the unified field-based perspective all things that exist, including any form of spacetime, come from the infinite eternal source of everything and remain within it. This is a holistic picture encompassing everything, not a reductive picture to black holes or literally *nothing*.

The contrast between a bottom line of nothing and the unified field as everything reflects a crucial change in perspective from reductionism to holism that fortunately has been building in recent decades. In reductive physical theory and modern science generally, the bottom line of nature is said to account completely for all processes that emerge from it. With respect to mind, this means that conscious mind is completely accounted for—*supervenes on*—processes in the brain/body, such as electrochemical neural activity. And with respect to cosmology, the bottom line is the consensus inflationary big bang theory that everything came from randomness and ultimately nothing. This is opposite of unified field theory, in which everything comes from the perfectly orderly unified field.

On the other hand, the basis of everything could be thought of as nothing in the sense of 'nothingness,' some abstract sense of existence so transcendent as to appear to be nothing. It can be viewed as nothing or everything, depending on the perspective preferred, whether reductive or holistic. But these different perspectives turn out to have quite significant practical implications. They may actually be a crucial factor in determining whether human life as we know it continues—discussed in the last chapter of this book.

The reductive argument for everything coming from nothing is exemplified here by summarizing recent work of physicist Lawrence Krauss.[36] Starting with the classical notion of space as empty nothing and then the quantum notion of it as vacuum energy, he goes further to posit everything coming from *literally* nothing. Consistent with discussions in this book, however, Krauss' description can be viewed as again revealing three important ingredients from which the physical emerges. What he refers to as *nothing* (no space, no time, no anything) happens to include quantum gravity as well as other ingredients.

In this model quantum gravity in the form of vacuum energy further is said to be 'unstable.' [36] But rather than to posit *nothing* as unstable, it seems more reasonable to say that 'something' has the inherent tendency to change. This relates to the notion of *inherent dynamism*, also associated in the model with the 'heat bath' in fundamental quantum processes.

Further, 'nothing' also includes symmetry and at least an initial balance of potential matter and antimatter. This balance of net zero energy became an asymmetry 'driven by some kind of random initial condition.' In 'nothing,' we now have random processes that comprise different states or initial conditions, which drive 'nothing' into physical asymmetry and our known universe based on fundamental laws of nature. Another way of saying this is that *inherent dynamism* inevitably is expressed in emergent natural laws and a universe relatively stable enough to exist, at least for awhile. This more rational view links 'nothing' with the three principles or 'forces' described in earlier models as fundamental to an emergent physical spacetime gravitational field: *inherent dynamism, inherent order, and inherent resistance to change*—energy, information, and matter or mass.

'Nothing' may not seem to include space, time, or anything but at least the potential for them as well as quantum gravity, processes resulting in laws of nature, and a relatively stable universe. Again, this certainly sounds like something, not 'nothing.'

What about God? The arguments by Krauss [36] for and against something emerging from nothing both seem to assume a reductive perspective. Paradoxically they place God, or the notion of a Creator or First Cause, outside of the universe. This can be understood in the context of Gödel's conclusion that mathematics, and any system of logic generally, involves assumptions that are outside in the sense of transcending the system and not provable within it. A more coherent view seems to be that totality is both the expressed aspect of nature (including the intellect and logic) and the unexpressed, which is transcendent and outside *from the view of the expressed.*

The point here is that the notions of God and the finite universe, infinite and finite, transcendent and immanent, need to be understood at the same time as an underlying unity. This corresponds to the approach that the reductive view of the finite relativistic spacetime gravitational field is embedded in a holistic view of spacetime as emerging ultimately within infinite eternal 'spacetime.'

In Krauss's [36] reductive view, there is physical reality and 'nothing,' or as we have discussed, more accurately 'nothingness.' This can be viewed as a two-level ontological model of the relativistic spacetime gravitational field and the unified field of nothingness. As discussed earlier and to be articulated more fully in Part II, this two-level ontology is not rich enough to account for mind, and doesn't incorporate progress to three-level models—the theme of Chapter 5.

Krauss also relates the contrast between everything and nothing to the major debate between secular and religious worldviews. Unfortunately these views are characterized in terms of theological versus scientific beliefs about the source of order in nature, and whether God as the Creator of the universe really exists.

Fortunately holistic unified field theory can be understood to bring back into the scientific picture core qualities that Newton, as well as most historical traditions, attributed to 'God.' Whatever else is attributed to God in various theistic views of nature, at least God as omnipresent, omnipotent, infinite, and eternal would have the same qualities of a completely unified field as the source and container of everything. In this holistic view 'God' is in each of us, at least in terms of omnipresence. The view of an external God as separate from creation, and from us as created beings, emphasizes other personal iconic attributions of God, without giving proper due to the totality of God as omnipresent, infinite and eternal—and also without grounding totality inside us, which is typical of reductive thinking.

With respect to cosmological theories, the all-encompassing holistic view of 'God' as the unified field does not translate into the big bang as an explosion but rather as an implosion within the infinitely extended unified field. Creation then would be in terms of the expression or emergence of more concrete levels and parts of nature, while remaining within unity. The whole creates the parts.

That again is a much more integrated and bigger picture, even fundamental enough to unite science and religion. And it has immense positive implications for a logically consistent model of nature and the place of human life in it that reductive perspectives fail to capture, and unfortunately argue against divisively.

The reductive view is commonly associated with scientific secularism. But arguments for the view rarely seem to address implications of the unified field perspective which is compatible with religious views of God as omnipresent. They also seem not to consider at all the recent important progress of theories toward nonlocal mind. These are key steps of progress consistent with both science and religion, important for bridging the gap between them and reducing the hostility both approaches have exacerbated.

The models described in the laudable progress toward unification in modern science tend to emphasize fundamental dualities not yet reconciled. Unified field theory can accomplish the long-sought goal of unification, at least on a theoretical level. But objectified modern science has not recognized the possibility of epistemological means for empirical validation of unity, the singular emphasis of the Vedic tradition. In the final chapter, we will briefly consider how the ancient Vedic approach includes systematic means for empirical validation of unity, with immense practical implications.

Toward this objective, it may again be of value to emphasize that the progress we have overviewed is moving toward a fundamental trinity of the creative-maintenance-dissolution operators, knower-process of knowing-known, and the consciousness-mind-matter ontology. This can as well be viewed as an expression of the fundamental principles of the abstract trinity in some major western religious traditions, and corresponding trinity in the Vedic tradition. Further, these can be viewed as directly related to emerging scientific theories of the physical level of nature, the non-physical level, and the transcendent unified field.

As discussed in this and preceding chapters, in the ancient Vedic model there also are three fundamental ingredients or principles from which emerge all the phenomenal objects and processes in finite nature. The three principles relate to the three-level model in terms of a gradient from the unified field to phenomenally interdependent subtle levels to the more diverse, independent gross levels—all remaining within the unified field. The concrete gross levels are associated with ordinary inert, relatively disorderly matter/energy and the more abstract subtle levels are associated with increasing orderly energy/intelligence. The principle of activation or energy has gross and subtle forms that extend across these levels, with less energy associated with the more expressed or grossest, inert physical level. This also directly can be seen in the progression in physics in the increasingly abstract focus of gross to subtler in term of mass to energy and now to information.

In reductive physicalism, the gross or surface level is the primary locus of experience and understanding. But contemporary progress overviewed in this book is pointing to a bigger picture that is logically consistent and more comprehensive, with the power to resolve long-standing dilemmas and paradoxes which physical and quantum theories have not achieved. Hopefully it is becoming clearer that physical and big bang theories are products of an engrained reductionism that overlooks a subtler sense of the entirety of nature as an infinite eternal holism or Oneness that includes everything.

In the next chapter we will overview recent three-level models that clearly exemplify progress toward the holistic Vedic three-in-one model of nature. This ancient model applies fundamental principles repeatedly referred in this book to describe dynamics and mechanics of the ontological structure of the finite universe. This is crumbling long-standing pillars that have held up the much more limited and inert physical theory, rendering it untenable and incomplete.

But in doing this, it also is replacing reductive physical theory with a much bigger and more consistent picture of the universe. It is not only much more integrated, but personally meaningful because it rationally includes us as real conscious beings in a much bigger, developmentally significant scientific picture. In other words, while maintaining either our scientific secular perspective or particular religious perspective, we also can progress on systematic spiritual development toward scientific knowledge and experience of unity.

Chapter 5

Three-Level Models

Physical theory presumed that mind is accounted for by brain processes and is nothing other than physical matter. But mind and mental processes have not been found in the local brain, though there certainly are increasingly extensive findings of neural processes in the brain that correlate with them. Reductive research to find the essence of matter has gone into finer-grained levels and into nonlocal quantum reality where mind might really exist, toward a consistent model of objective and subjective aspects of nature.

On the other hand, unified field models usually posit that the four fundamental forces begin to emerge at the Planck scale, held to be the smallest possible scale of ordinary space and time. This can be viewed as a two-level model, in which the universe built of the four fundamental forces is based in the unified field. Typical of the history of modern science, this two-level model does not explicitly account for mind. It has not been integrated with emerging theories of an ontologically real subtle field or implicate order underlying physical spacetime that is making room for mental space and real mind.

Three levels of nature within the totality. The more expanded models of a subtle non-physical field are progressing toward three ontological levels: the unified field, and within it the subtle nonlocal non-physical level and the gross local physical level. As described in earlier chapters, the prominent scientific theories most explicit in addressing where real mind fits into this bigger picture are loop quantum gravity theory that posits information space but not yet integrated with real mind, and neorealist quantum theory that much more explicitly posits real, causally efficacious individual minds in the nonlocal implicate order.

However, the neorealist interpretation is not psychophysical parallelism or dualism, because the levels causally interact through the psi wave. The explicate order emerges out of the implicate order, which requires that they causally interact. Finally in modern science we have a model ontologically rich enough to consider how real matter actually could link to real mind. Before this, we didn't have a model of mind as real, logically needed for it to interact causally with real matter and cause change in nature.

In the neorealist interpretation the local explicate order is generated from and permeated by the nonlocal implicate order. However, both the explicate order and the implicate order are within an ultimate universal plenum or *super-implicate order* akin to unified field theory. [17, 18] Thus the neorealist interpretation can be said to be a non-dual or monistic account of nature with three levels.

Clear correspondences can be identified between the neorealist model and other recent three-level models. For example, mathematician and cosmologist Roger Penrose [37] posits a three-level model, reflected in the quote below. The model attributes to abstract mathematical forms a real ontological existence apart from mental conceptions about them, in a very abstract *Platonic realm*:

"I am aware that there will still be many readers who find difficulty with assigning any kind of actual existence to mathematical structures. Let me make the request of such readers that they merely broaden their notion of what the term 'existence' can mean to them. The mathematical forms of Plato's world clearly do not have the same kind of existence as do ordinary physical objects such as tables and chairs... Objective mathematical notions must be thought of as timeless entities and are not to be regarded as being conjured into existence at the moment that they are first humanly perceived... Those designs were already 'in existence' since the beginning of time, in the potential timeless sense that they would necessarily be revealed precisely in the form that we perceive them today, no matter at what time or in what location some perceiving being might have chosen to examine them... Thus, mathematical existence is different from physical existence but also from an existence that is assigned by our mental perceptions. Yet there is a deep and mysterious connection with each of those other two forms of existence: the physical and the mental... I have schematically indicated all of these three forms of existence—the physical, the mental, and the Platonic mathematical—as entities belonging to three separate 'worlds'... There may be a sense in which the three worlds are not separate at all, but merely reflect, individually, aspects of a deeper truth about the world as a whole of which we have little conception at the present time." (pp. 17-23)

Another recent three-level model is proposed by physicist Henry Stapp. [38, 39] In the following quotes, three levels of nature are used to address the interaction of objective and subjective, matter and mind. Stapp [37] discusses the role of conscious mind within the orthodox interpretation of quantum wave function collapse. He asserts that consciousness is needed because:

"...the local-reductionistic laws of physics, regarded as a causal description of nature, are incomplete.... The physical part of reality represents merely the possibilities for an actual experience, not the actually experienced reality itself." (p. 213)

[F]rom the purely physical standpoint the [wave function] collapse seems to come from nowhere, as an unpredictable and undetermined 'bolt from the blue.' Something is needed to...bring 'classicality' into the dynamics, and it needs a 'cause' for the collapse, and it needs a reality to complement the 'potentia'... It must be something that exists, and the only thing that we know exists, besides the physical part of reality...is the experiential part...."

"...[T]he whole process is represented in the Hilbert space in which the quantum analogue of matter is represented. But rising out of the matter-like aspects of nature lies another dynamics governed by the experiential aspects of nature... What is important is the presence in the physical substrate of potentialities for quantum actualizations of experiential structures... *Is experience a fundamental element of nature, or is it derivative, or emergent?* It is fundamental because the fundamental realities are experiential." (p. 214)

Stapp's model includes physical reality, experiential reality, and Hilbert space. Hilbert space has a similar place as the universal plenum or super-implicate order in Bohm's model and also Penrose's 'Platonic realm.' It also has clear similarities with unified field theory, held to be the basis for the quantum and classical levels that are more comprehensively described in terms of nonlocal non-physical and local physical levels.

An additional three-level model, emphasizing the unified field-based perspective, has been proposed by unified field theorist John Hagelin, [40] even more consistent with the Vedic model of unity. This approach utilizes the abstract mathematical *Lagrangian* formulation.

In very compact form, the Lagrangian contains two terms. The first term, denoted as *phi*, can be described as a classical conception of a static space and time translation invariant field—a non-changing field of *existence*. The second term represents orderly dynamism or change, denoted as *II*. This term can be thought of as representing the inherent capability of the field to generate orderly change in the field. Hagelin associates it with the most abstract interpretation of the *quantum principle*:

> "To some extent, we can trace this property of intelligence to the fact that the unified field, beyond its mere existence, has a very precise and definite mathematical structure. This structure is typically defined in terms of symmetries of the field—invariance with respect to a set of internal and external transformation, such as Lorentz invariance, super-symmetry, modular invariance and gauge invariance…. The precise mathematical structure of the unified field serves as an unmanifest blueprint for the entire creation: all the laws of nature governing physics at every scale are just partial reflections or derivatives of this basic mathematical structure. However, this view of intelligence in terms of the classical symmetries of the unified field is a rather passive and inert one. The term 'intelligence' achieves its full significance only at the quantum-mechanical level of description, in which the field acquires a degree of *dynamism, discrimination and creativity* not present at the classical level." (p. 9)

The Lagrangian formulation places the unified quantum field in Hilbert space, a complex vector space of infinite dimensions—an infinite collection of points that comprise all states of a quantum mechanical system. Hagelin uses the Lagrangian formulation in

Hilbert space to present a unified field theory that more explicitly includes principles identifiable with the concepts of the knower or observer and the process of knowing—not merely the known. [40]

The *knower* or observer quality of the field is interpreted as the property of the Hilbert *space* of states to be a non-changing, unmanifest background for all possible unitary transformations or states of the field, while itself remaining completely unchanged. It is the uninvolved 'observer' of all transformations that, through its dynamic orderliness associated with the discriminative role of the inner product in evolving the quantum mechanical system, determines the physical manifestations of the system. The *process of knowing* quality of the field is related to quantum mechanical *observables* that serve as quantum mechanical operators in Hilbert space, generating changes of one state into another in unitary transformations. The *known* is interpreted as the stable quantum mechanical *states* themselves.

Thus this model incorporates in mathematical terms the three aspects of knower or observer, process of observing, and observed. The observer is related to Hilbert *space*, the process of observing to the quantum mechanical *observables*, and the observed to the quantum mechanical *states*.

In this mathematical representation, with the addition of dynamic order or intelligence associated with the quantum principle, any particle can be translated into any other particle. Even fermions and bosons—the most fundamental difference in particle theory—become unified and indistinguishable. This provides a unified field theory that in mathematical terms is matching up more completely with the unified field as universal Being in ancient Vedic literature.

In the quote earlier in this chapter, Penrose [12] noted that "we have little conception at the present time" about the worlds underlying the physical. Bohm made a similar comment in a quote in Chapter 3. But there is a well-articulated three-level model in ancient Vedic literature. [3, 40] The three-level models just described are progressing directly toward the Vedic *three-in-one* model. This model will now be introduced, and then elaborated in Part II.

Three-in-one Vedic model

The ancient Vedic tradition is sometimes described as the oldest continuous tradition of knowledge. Like modern science, it also pursues total knowledge of all the laws of nature. The word *Veda* can be translated as 'total knowledge.' [41] The closest concept in modern science to the ultimate totality, wholeness, or unity of Veda is the unified field as the source and container of everything. [5]

The *Sankhya* aspect of the Vedic literature describes a unified account of nature with three levels, which can be associated with gross physical, subtle non-physical, and all-encompassing transcendent levels. These levels are said ultimately to be nothing other than unity, the *three-in-one* unity of nature.

In the Sankhya enumeration, each manifest level of nature can be said to be a real phenomenal level with universal and individual aspects. Each level is a cosmic level that also includes corresponding levels and functions in individual human beings. [42]

Thus universal Being or universal consciousness has within it individual being or individual consciousness—another expression of point/infinity or part/whole at the same time. And cosmic ego, intellect, and mind have within them corresponding individual levels. Soon we will show how this model of levels of mind or the inner dimension has correspondence with the model of mind developing in scientific psychology over the past 150 years. [5, 33]

From within the infinite eternal unified field, finite phenomenal levels are expressed with increasing limitations from subtlest to grossest. Individual minds and individual objects of sense—the subject-object duality—also emerge with the increasing limitations from subtlest to grossest. Each individual is comprised of all levels in degrees of expressed or latent form.

The following chart is proposed as a reasonable interpretation of the Sankhya model. It serves as a framework to show how the three-in-one Vedic model is the direction of progress of the three-level ontological models of nature just described. It also serves as a framework for the detailed discussions linking mind and matter in Part II, and will be referred back to from time to time as we proceed in upcoming discussions.

-------------------------------Gross Physical Level-------------------------------

Gross sensory environment	Gross sensory environment
(Infinity to ultra-macroscopic)	
(ultra-macroscopic to ~10^{-3} cm)	
Brain/body	
Electrochemical cellular processes	
(Microscopic ~10^{-4} to~10^{-8} cm)	
Atomic/ sub-atomic processes	
(Ultra-microscopic ~10^{-9} to ~10^{-32} cm)	
Quantum field/ quantum gravity	
Planck scale (~10^{-33} cm)	
	Mahabhutas

-------------------------Subtle Non-Physical Level-------------------------

	Subtle Sensory Environment
Nonlocal information space	
Nonlocal, non-quantized space	
	Tanmatras
	Sense Organs Action Organs
	Gyanendriyas Karmendriyas
Mind (thinking)	Manas
Intellect (discrimination and	Ahamkara
feeling)	
Individual self or ego	Mahat

-----------------------Unmanifest/Transcendent Level---------------------

(Infinitesimal point/ Infinity)	
Unified field	Prakriti
	Universal Being/Purusha

Figure 1. This chart speculates on direct comparisons of how three-level quantum models relate to the Sankhya model, with the gross local level permeated by the subtle nonlocal level, both ultimately nothing other than the transcendent unified field or universal Being. The left column presents an interpretation of the three-level models with the addition of a scale of spatial dimensions speculating about their relationships. The right column presents an interpretation of the levels in Sankhya, using Vedic terms (Each Vedic term will be given more detailed description in Part II). This chart is a vertical depiction of grosser levels embedded in and encompassed by subtler levels, all within the unified field of universal Being.

The above chart represents in a linear graphic display a completely holistic view in which all the parts of nature appear to emerge from, while at the same time remaining within, the unified field. The first levels are the most similar to the unified field in that they have the fewest limitations. Phenomenal manifestation progresses in sequence from subtle levels to inert gross physical levels such as particles which compose elements, rocks, and earth.

Infinity in a point. A helpful strategy for understanding this completely holistic model is to disembed from the fragmenting reductive perspective that brings everything down to smaller and smaller scales. Instead of the universe narrowing down to an infinitesimal point such as a black hole or even nothing, phenomenal manifestation is in terms of limitations of infinity into finite values.

In other words, the whole precedes the parts and the parts come from it. The whole is not just a collection of the parts, and not just more than the sum of the parts. Space is not empty, and it and everything in nature does not come from nothing. All phenomenal levels are limitations of the infinite eternal unified field that is already everywhere. The infinite unified field contains within it all the expressed, manifest, finite levels of nature.

The level of ordinary conventional space within which our physical bodies move is the most abstract of the gross physical level. We have long had the notion of space as empty with only occasional pieces of matter in it, spread out through the vast visible cosmos. But in the Vedic model, this level is only the crusty surface of a much vaster phenomenal universe. Inside is much bigger than outside; indeed, outside is finite and inside is infinite.

Scientific progress has led to subtler nonconventional space and even mental space, much more abstract and extended than the gross physical level. And these fields are derived from the unified field that may appear as if it is nothing but contains everything. It can be said to be 'nothingness,' but not literally nothing. In the Vedic model, the ultimate unified field of nature is the only infinite eternal reality. As Maharishi [43] has emphasized:

"Unity is real. Diversity is conceptual."

Higher dimensions? We have now overviewed progress toward the ontological reality of levels of nature beyond the most tangible physical level. The progress is leading further to recognition of nonlocal levels including mind. But these are relative, finite phenomenal realities in light of the infinite eternal unified field.

From the holistic perspective, there may be no need to posit higher-order dimensions beyond four-dimensional spacetime to account for subtle nonlocal processes beyond the Planck scale. The difference between subtle nonconventional spacetime and gross conventional spacetime may not be any new higher-order or hidden spatial dimensions, given the holistic view that they are limitations within the infinite eternal unified field. Nonconventional space, higher dimensions, or the hidden sector can be viewed as attempts to envision subtler levels in the context of reductive physicalism, local causality, and mathematical concepts not yet grounded in holism.

In the more expanded context that is emerging, theories that emphasize Planck-size quantization can be considered basically corpuscular models of smaller time and distance scales than their atomistic predecessors. In these theories, strings, loops, branes, qubits of information—however the smallest entity, process, or event is envisioned—embody some notion of a membrane or boundary that delimits them. This again brings up the issue of infinite regress.

The Planck scale is thought to be the smallest possible size of an object and boundary between objects. But at least theoretically, the boundaries of the object could be thinner and thinner. On the other hand, there needs to be some level at which discontinuous quanta merge into indivisible continuity if there is a *completely* unified field—a field beyond all gaps and boundaries, beyond all differences, completely unitary and one with itself. If there are boundaries in terms of Planck-unit discreteness, then it is not completely unified.

The quantum principle in terms of the Planck scale ultimately cannot be fundamental if there is a completely unified field. Whether this transition to unity takes place at the Planck scale or at some even subtler level that permeates the Planck scale is quite significant. The resolution to this issue requires an expanded conception beyond relativistic and quantum notions of spacetime.

Planck unit quantization is the compactification? Contemporary models posit unification as at the level where the fundamental forces merge into a single field at the Planck scale—the hypothesized level of super-unification, the field of supergravity or quantum gravity. This can be viewed as providing a model of a unified basis for all material objects, but not of a *completely* unified field. If the unified field were to be undifferentiated in its unity, it would have to underlie any discrete Planck-unit quantized field, including quantum gravity in terms of the Planck scale as a physical dimension.

Abstracting matter into an extra-dimensional conceptual or quantized information space, described as a pure geometry of discrete information qubits from which ordinary spacetime is generated— toward a 'minddust' theory—is more abstract and thus a step closer. But it still doesn't yet describe a *completely* unified field.

In the holistic view, nonconventional space associated with information and mind can be placed in-between ordinary local spacetime and the supersymmetric unified field. This subtle field, for lack of better terms, could be described as both smaller and bigger than physical existence in the sense of permeating it, but still finite. It would be hidden with respect to ordinary or conventional spacetime, but not as higher-order compactified spatial dimensions. It would be quantized, but not in Planck units of space.

Rather, it would be *unfurled* and much more extended than relativistic or quantized spacetime. It would be hidden in a somewhat comparable way in which space is hidden on the physical level of sensory experience. The more abstract, subtler notion of a non-physical, nonlocal ethereal (not empty) field permeates the more restricted local spacetime of the gross level of nature in our ordinary physical world. Reductively space is *compactified* at the Planck scale. In string theory the classical levels are where space is unfurled, not enfolded, and spatial dimensions near the Planck scale are enfolded, not unfurled. Here the opposite view is proposed. [5]

Quantization in physical Planck units may be the limiting of an even more abstract, underlying, unfurled, nonlocal field into discrete, localized enfolded Planck-unit quantum waves that are further expressed into discrete particles. It thus could be viewed that Planck scale quantization *is* the compactification. Some properties of space associated with the gravitational field are quantized. These are expressed in the process of sequential symmetry breaking when Planck unit quantization takes place and nature limits itself to matter particles—associated with quantized space as in quantum gravity.

This may occur for example at the level of symmetry breaking in which the nonlocal background field emerges into the gravitational and strong nuclear force fields, hypothesized to occur at the Planck scale. Physical objects built of particles are subject to the limitations that characterize the field of quantum gravity and that exist at the level of the Planck scale and larger. This includes all physical objects.

But it might not include subtler processes not quantized in Planck units and not physical. These subtler processes would be nonlocal processes not subject to the limitations of the relativistic spacetime gravitational field characterized by local causality and particle interaction mechanics which generally relate to the Planck scale. If nonlocal causal dynamics are not limited by local causality and light-speed, then they would need to be beyond or subtler than the relativistic spacetime gravitational field.

Much of the difficulty of envisioning a nonlocal level may be due to the strong reductive tendency that characterizes modern science. It is not surprising given this tendency that models propose a smallest unit of spacetime and closed physical causal nexus as the ultimate level of nature. The habit of trying to understand nature based on *idealized* mathematical concepts without considering what is ontologically real may add to the difficulty. It may have led to hypothetical constructs to account for the lack of fit between physical levels and the non-physical level closer to a dimensionless point.

In this context the indeterminacy associated with hypothesized quantum fluctuations at the Planck scale in part may be due to applying a dimensionless point mathematical model to the Planck unit quantum level. Superimposed on particle and wave properties of objects are idealized mathematical concepts based on the model of a particle as a dimensionless point with no extension in space. To relate this mathematical model with the particle and wave properties, it has been theorized that mathematical points fluctuate randomly—

vacuum fluctuations—and thus limit measurement precision to an average proportional to ½ of Planck's constant. At least one source of the indeterminacy of nature may be related to the mathematical model of *dimensionless* points superimposed onto real quantized Planck-size processes—and onto wave dynamics.

Inherent dynamism may be expressed at the quantum level in terms of vibrations or pulsations of Planck unit quanta. But some dynamics of inherent dynamism may be related to the subtler non-quantized level of nature, also not involving thermodynamics—that is, more associated with an abstract information field and mental thought waves which are not 'hot' but rather are more abstract energy forms. In this view, though mind and consciousness are at the basis of the physical, they are much subtler than it and are not 'hot' in the sense that is attributed to a black hole singularity near about the Planck scale, for example.

A dimensionless point is not quantized in the sense of Planck unit extensions of space and time. It can be viewed as a way to account for discrete particle mechanics that are more fundamentally real wave dynamics that don't fit the local particle interaction causal model. A dimensionless point is a subtler concept than either particle or wave aspects of the quantum. It is an intellectual division superimposed on an infinite space, with no extension in space. This is a more abstract notion of the *quantum principle* than the quantized level of the Planck scale, more in line with the subtler concept of this principle with respect to the notion of discriminative intelligence described by Hagelin [3] in Chapter 3.

Further, the reductive approach has the impossible task of explaining how everything comes from nothing, and how the whole emerges from the parts into more than the sum of the parts that then causally influences the parts. The holistic view has the opposite and more logically consistent task of explaining how parts emerge as limitations of the whole.

The unified field perspective begins with ultimate unity: the whole creates the parts, and the parts remain within the whole. The ultimate wholeness is simultaneously *smaller than the smallest and bigger than the biggest* (Katha Upanishad 1.2.20), [44] beyond ultimate reductionism and ultimate holism. It thus can be described as being prior to any intellectual conception of part *or* whole, or even part *and* whole.

All finite ontological levels, with all the vast diversity in nature, would somehow have to be within the unified field because there is nothing outside it. It would have to be at the same time infinitely diverse and infinitely unified, again beyond diversity and unity.

As hinted at in Chapter 1, this model can be said to turn nature downside up. The bottom line contains everything, not literally nothing. Also, the subjective mind is more fundamental than objective matter. Matter emerges from deeper levels of mind and consciousness, not that conscious mind is a product of the brain and underlain by inert, random bits of matter/energy. In that view, the unified field ultimately would have in it the quality of consciousness itself.

Direct knowledge of totality, incorporating both diversity and unity at the same time, requires going beyond the discriminative intellect. How to do that is discussed in Part III. But next, in Part II we will go into more specifics about the theme of a consistent model linking mind and matter based on the three-in-one ontology of the completely holistic Vedic approach.

PART II

LINKING MIND AND MATTER

Chapter 6

Eight-Fold Structure

For centuries the philosophical depth of the ancient Vedic tradition was recognized among many scholars, but its practical value was not applied or properly understood. In recent years Vedic proponent and educator Maharishi Mahesh Yogi has reestablished its completely holistic value and has focused on reviving its practical applications, as *Maharishi Vedic Science and Technology*.

The Vedic account conceives nature as manifesting in the manner contained in the 10 Mandalas of the Rik Veda, which is the first chapter of the Veda. It is said to encompass the entire phenomenal ontology of nature. Within the ultimate wholeness, the first and tenth mandalas focus on unity and transcendent non-duality. The other eight Mandalas describe the *eight-fold* structure of the phenomenal universe, each a whole in itself. It covers three fundamental qualities or forces within the non-dual unity, the three-in-one self-interacting dynamics that condenses into five fundamental constituents. This view is broadly consistent with sequential symmetry breaking, the 'arrow of time,' and the second law of thermodynamics that imply the universe emerged from the lowest entropy, super-symmetric, all-encompassing unified field. In this view space is derived from infinity, time from eternity, and mortality from immortality.

Sankhya is an aspect of Vedic literature called the *Darshanas*. This aspect includes an ontological analysis of levels of nature that will now be shown to be particularly relevant to the theories we have been reviewing. It is important in examining Sankhya, however, to keep in mind the other Darshanas, especially *Vedanta*. In that perspective all phenomenal levels are ultimately nothing other than indivisible unity.

We will first draw correspondences between fundamental forces of nature in the ancient Vedic and modern scientific traditions. We will then examine in more detail the levels of mind in the subtle non-physical level, in order to establish the basis for addressing how mind links to matter in the gross physical level of the brain/body.

Three ingredients, principles, qualities, or 'forces'

In any conceptual delineation there is an initial duality, basic to how the discriminating intellect functions. And in duality there is an implied trinity, concerning the relationship between the dual aspects. Concepts of the trinity can be found in many cultural traditions. Most ancient sciences including the Vedic tradition in India and elsewhere around the world share with modern science a simple binary logic from which emerges three-fold and sometimes four-fold and five-fold models, in light of various schemes akin to symmetry and laws of conservation in modern physics. [45]

For example, in the delineation of observer and observed there is the connecting process of observing; in subject and object there is the predicate, In Father and Son there is the Holy Spirit. Also, in Vedic literature there are the fundamental trinities of knower-process of knowing-known, as well as Brahma-Vishnu-Siva rishi-devata-chhandas on all three levels, sattva-rajas-tamas on the subtle and gross levels, and vata-pita-kapha on the gross level associated with the physical body.

In the ancient Vedic tradition the three abstract principles relate to both the nature of the discriminating processes of the intellect and the structuring dynamics of all phenomenal objects in nature—both subject and object, both the structure of the observer and of the objects observed.

In Sankhya three qualities or forces materialize five constituents. This can be quite helpful for integrating contemporary particle-force theories positing a multitude of particles emerging from the four quantum fields which gain mass via the Higgs force particle.

The three forces identified in Sankhya and Vedic literature are frequently called *sattva, rajas*, and *tamas*. These three qualities are described as inseparable, co-existing, and co-functioning in relative degrees to carry out every interaction at all phenomenal levels of existence. They shape infinite potentiality into relative finite nonlocal and local phenomenal actualities. They also can be related to the three aspects of time—future, present, past—the three spatial dimensions—x, y, z axes or up/down, forward/backward, and right/left—and other trinities. Although entangled and *self-interacting*, in simple terms they relate to the basic creative (propagation), maintenance, and dissolution operators. In the following quote, Maharishi [42] elaborates on sattva, rajas, and tamas as the *gunas* or fundamental principles, qualities, ingredients, or forces of nature:

> "The entire creation is the interplay of the three gunas. When the primal equilibrium of sattva, rajas and tamas is disturbed, they begin to interact and creation begins. All three must be present in every aspect of creation because, with creation, the process of evolution begins and this needs two forces opposed to each other and one that is complementary to both. Sattva and tamas are opposed to each other, while rajas is the force complementary to both. Tamas destroys the created state; Sattva creates a new state while the first is being destroyed. In this way, through the simultaneous processes of creation and destruction the process of evolution is carried on. The force of Rajas plays a necessary but neutral part in creation and destruction; it maintains a bond between the forces of sattva and tamas (pp. 269-270)."

On the gross local physical level, these fundamental qualities or forces can be related to the principles of attraction (gravitation), activity (inherent dynamism), and inertia or resistance to change (mass, Higgs fields). These principles are associated with corresponding principles discussed in Part I as important ingredients of contemporary theories in physics. But these principles or qualities have not yet been integrated into the model of four fundamental forces of nature and Higgs field theory.

Sattva. This principle, quality, or force can be associated with the maintenance operator, upholding and fostering balanced change and continuity. It is the unifying principle—the attraction, balancing, or harmonizing value of nature. In the physical universe it can be associated most directly with gravity, attraction to the center point of an object, and the gravitational constant. It also can be related to the third law of thermodynamics and negentropic processes in nature. It refers to decreased activity associated with decreased temperature in material systems which results in decreased entropy, a fundamental order-increasing or *negentropic* process of meaningful information that can be associated with the maintenance of order throughout phenomenal nature. It is in this sense that it can be seen as relating more to the principle of 'Vishnu' in ancient Vedic literature.

In the subtle relative domain or level, sattva can be said to unify oneness toward Oneness, individuality toward universality; individual point value or individual self toward universal wholeness or universal Self. With respect to levels of individual subjectivity, it can be understood to maintain the unitary sense of individual self—centering on the individual self or ego that through evolution unites with the universal Self. At the gross physical level, it can be conceptualized as unifying into a dimensionless point value and elementary particles— the attractive force of gravity toward the center point of an object.

Thus the gravitational or attractive force that is ubiquitous in the physical universe is a gross expression of a much more abstract principle of the attractive or unifying principle of sattva. Quite importantly, this principle also can be related to the notion of attraction on the subtle subjective level, traditionally associated with the attractive, unifying principle of love.

In this regard, Einstein made the comment that, "Gravity cannot be held responsible for people falling in love." [46] This point is certainly valid within the restricted gross physical meaning of the concept of gravity. But in the completely integrated Vedic approach, it can be viewed as the same universal principle or force of attraction manifesting on different levels of phenomenal nature.

On the gross level it is the concept of gravity precisely characterized in Einstein's formulation. But on the subtler, more abstract subjective level it can be associated with the attraction of individual love, and ultimately toward universal Love or Unity.

Thus the principle extends beyond Einstein locality into the subtle relative nonlocal field of which Einstein seems to have had little inkling. It is in this integrated sense that the following statement by Maharishi [47] can be placed:

"The force of gravity will be ultimately found to be the unified field administering the whole universe."

Rajas. This principle can be related to inherent dynamism, associated with the propagation or creation operator, activating the maintenance and dissolution operators. It provides neutral energy that impels change. In the physical it can be associated with energy and propagation, expressive or diversifying processes following laws of energy conservation, and perhaps light-speed and Planck energy.

Rajas also can be related to the first law of thermodynamics, which is that all motion or interaction between physical objects results in no loss in the total amount of energy. In the sense of the creative operator, it seems to be related more to the principle associated with the concept of *'Brahma'* in ancient Vedic literature.

Tamas. This third principle or force can be related to resistance to change or inertia, most closely associated with the dissolution operator which restrains the creative and maintenance operators. In the gross physical universe it can be related to the concept of mass and matter, Higgs field theory, and possibly Planck's constant.

In some sense tamas can be related to the second law of thermodynamics, which refers to the tendency in nature toward greater disorder and entropy. In this sense it leads to the least amount of change, the most inert states of nature in which the most localized point value predominates and in which the infinite dynamism is most hidden. Higgs field theory as a concept to account for particle mass seems similar to the more basic quality of tamas on the gross level.

It also can be taken further to be toward changelessness, toward hiding or becoming unmanifest—ultimately to eternal non-change and infinite silence. It is in this sense that this quality can be related to the Vedic principle of *'Shiva,'* representing infinite eternal silence.

The three values from which the hypothesized most fundamental unit of conventional spacetime, the Planck unit (10^{-33} cm), is derived—gravitational constant, light-speed, Planck's constant—may correspond with sattva, rajas, and tamas on the gross explicate level.[5] To apply these three forces to the ordinary physical level, they can be thought of as inherent in every point in an unbounded field. Thinking of an abstract field as being made of infinity of points, if each point has a certain property then the field also has the property, which gives the field overall textural qualities and a defining fabric. This may give a sense of how the quantitative values of the Planck scale, light-speed, and relativistic gravity relate to textural qualities of the fabric of the conventional spacetime gravitational field.

Additional Metric for the inner dimension. From the view of conventional space and time, the basic *metric* for measurement is distance such as in centimeters, and duration such as in seconds. Einstein's relativistic theories revealed the interrelationship of space and time, distance and duration. In these theories light-speed is an absolute value that is a defining limit of all interactions in physical existence. It characterizes the gross level of nature, the explicate order or relativistic spacetime gravitational field.

From the perspective of subtler levels of spacetime or levels of existence, a fundamental distinguishing metric related to the inner dimension can be thought of more in terms of adding the degree of interconnectedness or entanglement—the expressed degree of self-referral of infinity and point, whole and part, unity and diversity, interdependence and independence, universality and individuality, or nonlocality and locality. In the Vedic view this can be related to increasing sattva, the attraction and unifying quality that is more prominent at deeper, subtler levels toward ultimate unification.

In the transcendent completely unified domain there is *infinite* self-referral—Self-referral—and in the subtle and gross relative domains there are finite relative degrees of the expression of self-referral. In the subtle relative domain it can be thought of as a more limited referral to individual self, and in the gross relative domain it can be thought of as the most limited into a point particle. At the transition to the ordinary gross level of the relativistic spacetime gravitational field, self-interacting dynamics emerge in the form of spherical pulsations of energy such as atoms, planets, stars, or drops of water for example. This also can be related to the ordinary waking state experience of object-referral and subject-object duality.

According to this model, every 'point' in finite nature contains within it, or is built out of, these three ingredients, principles, or forces which carry out or implement all change. The forces can be viewed in terms of an outward, expansive or propagation force (rajas), a restrictive or inertial force (tamas), and a unifying or attractive force that maintains order between these other two forces (sattva). All phenomena in nature are due to relative degrees of these three forces. At the subtlest level closest to the unified field level, the unifying force of sattva is most predominant. As the restrictive or inertial force increases, nature is expressed in increasing limitations and diversity to the gross levels of inert matter particles. All phenomenal change is within the infinite eternal silence, implemented by these three forces.

Five constituents

In Sankhya the three gunas, principles, or forces materialize into five abstract fields or constituents of nature. On the gross local physical level, these five constituents are called the *mahabhutas*. The Vedic term mahabhuta is from *maha* (great, universal), *bhu* (curving back, giving form, to happen, occur, exist; *bhut* (creation), and *ta* (finished, created). [48] It relates to the ordinary local physical level of existence we are familiar with through our ordinary five senses in the ordinary waking state of human consciousness.

The mahabhutas can be viewed as 'great' or expanded fields or mediums that become quantized via curving back on themselves to express concrete limited properties in atomic particles with increasing relative degrees of tamas. They relate to the gross level of nature, in which the point or part value is most prominent and the infinity or wholeness value is most hidden due to the influence of tamas.

These five mahabhutas can be described as vibrations of the unified field or universal Being in its grossest, most concrete localized expressions. They also are associated with the classical concepts of the basic constituents of space, air, fire, water, and earth.

But unfortunately these terms have long been interpreted in a simplistic and misleading manner as a pre-scientific view of nature. A deeper analysis suggests that the mahabhutas refer to abstract processes that structure physical objects with the properties of vacuity (space), mobility (air), luminosity (fire), liquidity (water), and solidity (earth).

For example the mahabhuta of *air* not only refers to what we ordinarily think of as air, but also to abstract principles that manifest as gaseous processes, and also agglomerations into matter. The nature of the mahabhutas as abstract processes may be more obvious with respect to fire. The concept of fire as a fundamental constituent clearly suggests the abstract functional nature of the mahabhutas. It refers to the laws of nature in transformations such as radiation, combustion, oxidation, and illumination—not just a superficial description of basic elements attributed to primitive cultures.

The five mahabhutas make up the entire gross relative creation, comprising the ultramicroscopic, microscopic, macroscopic, and ultramacroscopic levels in the physical sciences within the local relativistic spacetime gravitational field. Each mahabhuta can be viewed as condensing (with more dense or localized structure) from the preceding one, and manifests an additional limitation or property, along with general properties of the others. The mahabhutas combine in innumerable permutations to create the vast diversity of the physical universe; but not creating new ontological levels.

In Vedic literature the five subtle tanmatras and the five gross mahabhutas can be described as being expressed in sequential symmetry breaking, each containing the qualities of the others but *expressing* most prominently its own specific quality. For example, the sequential interdependence of the five mahabhutas is sometimes described in terms of relative percentages of each one. The mahabhuta of space is described as composed 1/2 of the space quality and 1/8 each of the other four, the mahabhuta of air as 1/2 with 1/8 each of others, and so on from subtler to grosser. [49]

As physical realities of the ordinary phenomenal world, the five mahabhutas must in some way correspond to the quantized particle-forces in physics. The current state of knowledge may not be quite developed enough to establish the precise correspondences of the known particle-forces with the mahabhutas. But if both approaches refer to the same physical world, it is reasonable to expect eventually they would match up as our knowledge advances.

To link this system to the fundamental forces in physics, one reasonable speculation is that the mahabhuta of space is most closely associated with the gravitational force. Likewise the mahabhuta of air might express the gravitational and strong nuclear forces. The mahabhuta of fire might express the gravitational, strong, and weak forces. In this speculative comparison, the mahabhutas of water and earth might express all four forces, most associated with electromagnetism, and further electricity more prominent with water and magnetism with earth.

Space (Akasha). From the root 'to appear,' *akasha* relates to the abstract principle of *vacuity*, and seems to be most akin to the concept of the relativistic spacetime gravitational field. Every physical object is permeated by and shaped from the mahabhuta of akasha.

In modern physics, objects existing in this level have the limitation of light-speed. The level of akasha can be related to zero point energy, the Heisenberg uncertainty principle, local causality and the light cone, Einstein's relativistic model of gravity, particle interaction causality, Planck-size quantization associated with relativistic spacetime gravity, and the principles of thermodynamics.

The mahabhutas are sometimes described as dimensionless points, in the same sense as the mathematical point-particle concept used in calculations of motion in non-relativistic and relativistic classical physics. The mahabhuta of akasha or space is not described as having a particulate structure in the sense of quantum theories which posit spacetime as fundamentally discrete Planck-unit quanta mediated by a particle such as the hypothesized graviton.

However, the principle of vacuity of akasha is also sometimes conceived as having an additional quality of *porosity*, [48] which may correspond to these conceptions, as well as to spacetime foam. Although the general theory of relativity described space as relational, it nonetheless later has been associated with specific textural properties. It is understood further as not empty and not just due to relationships between physical objects or material bodies, which of course need to be in some form of space to exist.

In *Vaishesika*, another aspect of ancient Vedic science, the four mahabhutas other than akasha are identified as *paramanus*, sometimes interpreted as the smallest possible divisions of matter. The four paramanus (air, fire, water, earth) are characterized as having extension and magnitude in akasha or space, and thus can be associated with quantization and particle properties. [48]

One way to envision the physical world as Planck-size quanta is that when each point in a field has a quality of attraction or gravity, pulling toward itself from all directions—so to speak—and points in the field are differentiated or separate from each other in some conceptual and phenomenal sense, then the points would pull on each other. When the pull of each point with the points adjacent to it is from all directions and practically speaking of infinite extent, then a point on one side would pull in the opposite direction of the point on the other side, in all directions of the field. The opposing pulls would appear to establish each point as a specific functional point within the undivided field.

The mathematical concept of a point could be thought of as becoming quantized with physical extension, determined by the strength of attraction and the other counteracting forces of inherent dynamism and resistance to change. This would give a texture to the field, with the relative quantities determining the size of the quantum unit—the Planck size in the gross physical field.

In this example the spacetime field inherently embodies order, dynamism, and resistance to change, which can be associated with the abstract forces of sattva, rajas, and tamas. As tamas and its quality of inertia or resistance to change increases, the abstract field becomes more concrete, with localized objects in it made of the textural content of the field, expressing more specific properties characterized by localization and independence of objects such as the known particles. This would also define the limiting properties of the field, including for example the maximum speed of objects built from it that appear to move in it.

In the gross spacetime field of the physical we are familiar with via the ordinary senses, Newton's Gravitational constant seems to be directly related to sattva and the attractive force. Influences counteracting the force of gravity seem to be directly related to the natural numerical values associated with light-speed and Planck's constant.

The particle aspect of the field is said to be inherently dynamic, quantified in terms of Planck units and light-speed. It seems reasonable that this is directly related to rajas, activation, and the concept of inherent dynamism. Correspondingly the property of viscosity, inertia, or resistance to change in Higgs field theory seems related to tamas, and possibly the numeric value of Planck's constant. The three gunas or qualities thus may relate to the fundamental forces in modern physics. Their natural numeric values define Planck-size quanta and the limitation of light-speed, which comprise all physical objects in the ordinary conventional relativistic spacetime gravitational field.

Within this field, the five mahabhutas or gross constituents can be thought of as progressive limitations, each more expressed one embedded in the previous one. They also can be thought of as progressive layers in gross spacetime, each adding a specific quality expressed in different sensory phenomena.

A way to think about the paramanus is that they are structured by further limiting of the spacetime gravitational field—more sharply collecting into or *curving back onto itself*, into discrete concrete forms that function as relatively more independent particles. In this view the space mahabhuta expresses the force or principle of gravity.

This also is relevant to the contemporary model of space as 'flat' in the sense of infinite, extending in all three directions without being curved, which Greene [10] describes as the front-running contender for the overall shape of the universe. The notion of spacetime as flat might be associated with the unmanifest infinite level or transcendent level of spacetime, not just the coarse-grained physical level of Newtonian space and time. But with respect to finite space in relative creation, space can be thought of as curved. The notion of the curvature of space—such as into a sphere if conceived as curving back on itself—relates to finite limitation of infinite Self-referral.

To explain finite creation in the Vedic model, infinity is said to curve back onto itself, *infinite Self-referral*. [50] This curving back onto itself can be associated on the subtle finite manifest level with a *mandala* form. This is related to the Vedic concept of *Hiranya garbha*, [51] sometimes called the cosmic egg, the manifest form of the unified field curving back onto itself in the creation of the cosmic expanse of relative finite spacetime. On the level of conventional spacetime, the dynamic of curving back onto itself can be associated with a point particle, Planck-size quantum, atomic particles, and even magnets in which the lines of force curve back into the form—as well as the overall shape of the relativistic spacetime gravitational field and its 'negative' curvature.

Air (Vayu). From the root 'to blow,' vayu is related to the abstract principle of *mobility* or motion, and the related functions of pressure and impact, compression and rarefaction, most akin to the concept of *air*. The mahabhuta of air further condenses from the mahabhuta of space, which contains the principle of air.

In the increasing limitation of space, it is the nature of the gravitational unifying force to attract points of spacetime together into clumps or regions of more and less compression, which further precipitates into a gaseous state. The mahabhuta of *air* fills the available three-dimensional space—within the constraints of gravity—but has the additional limitation of not being able to permeate objects, which are properties of a gas.

With respect to particle-forces, the fundamental force that binds or glues particles into atomic nuclei and compounds is the strong nuclear force. In this speculative view the mahabhuta of air would express the gravitational force along with the strong nuclear force (again including the weak and electromagnetic forces, but more latent and not as prominent).

Fire (Tejas). From the root 'to be sharp,' tejas relates to abstract principles of *luminosity, form,* and *transformation*, associated with the fundamental element of fire. The mahabhuta of fire involves heat and temperature as well as radiation, combustion, and oxidation. Fundamental to *fire* is oxygen, a core element in combustion.

When there are aggregates of points as volumes in spacetime that cannot penetrate each other, like air, their agitation increases when further limited; pressure and activity rise, related to temperature or heat. At certain high temperatures, particles can be emitted in the form of kinetic energy, resulting in radiation, heat and luminance.

Continuing the comparison, it would seem that the mahabhuta of fire thus might relate to gravitational, strong nuclear, and especially weak nuclear forces that concern transformations in nature. Greene [10] makes relevant points in the following quote:

"Gravity is a universally attractive force; hence, if you have a large enough mass of gas, every region of gas will pull on every other and this will cause the gas to fragment into clumps... Even though the clumps appear to be more ordered than the initially diffuse gas—in calculating entropy you need to tally up the contributions from *all* sources.... For the initially diffuse gas cloud, you find that the entropy decrease through the formation of orderly clumps is more than compensated by the heat generated as the gas compresses, and, ultimately, by the enormous amount of heat and light released when nuclear processes begin to take place." (p. 172)

Water (Apas). Apas relates to the abstract principle of *liquidity* or *fluidity*. It has the freedom of flow or movement to fill the available space within the limitations of its permeability; but because of its lower kinetic energy and higher mass, only sort of 'downward' gravitational pull due to increased mass.

The liquid state, as in water, has additional limitations over fire, air, and space. There is less internal motion, less heat, and restriction of flow rather than gaseous expansion. Again Greene [10] discusses relevant points, with respect to symmetry:

"On a molecular scale, for instance, ice has a crystalline form of H_2O molecules arranged in an ordered, hexagonal lattice... The overall pattern of the ice molecules is left unchanged only by certain special manipulations, such as rotations in units of 60 degrees about particular axes of the hexagonal arrangement. By contrast, when we heat ice, the crystalline arrangement melts into a jumbled, uniform clump of molecules—liquid water—that remains unchanged under rotations by any angle, about any axis. So, by heating ice and causing it to go through a solid-to-liquid phase transition, we have made it more symmetric... Similarly, if we heat liquid water and it turns into gaseous steam, the phase transition also results in an increase in symmetry. In a clump of water, the individual H_2O molecules are, on average, packed together with the hydrogen side of one molecule next to the oxygen side of its neighbor. If you were to rotate one or another molecule in a clump it would noticeably disrupt the molecular pattern. But when the water boils and turns into steam, the molecules flit here and there freely; there is no longer any pattern to the orientations of the H_2O molecule and hence, were you to rotate a molecule or group of molecules, the gas would look the same. Thus, just as the ice-to-water transition results in an increase in symmetry, the water-to-steam transition does so as well." (p. 253)

Liquidity embodies the concept of flow—movement of impulses of energy through or along a specific path, such as a current of water in a river, a current of electricity, or a neural impulse. With respect to fundamental particle-forces, this would seem to be most closely associated with the electromagnetic force.

The outer shell of charged atoms allows electrons to flow, such as current through a medium of copper wire, from negative to positive and positive to negative electric charge. Electric current flows easily when electrons are loosely held; mediums that hold electrons more tightly are insulators. In this comparison the mahabhuta of water would express properties of all four fundamental forces, but most specifically the electromagnetic force, with emphasis on the flow quality of electricity.

As noted in the Introduction, historically electricity and magnetism were two forces, before their underlying symmetry was recognized. This symmetry so intimately connects them that they are not characterized as differentiating through symmetry breaking in the same way as the other forces. However, physical objects can appear to exhibit electricity or magnetism, as well as both or neither.

Earth (Prthivi). From the root 'broad or extended,' the mahabhuta of prthivi relates to the abstract principle of *solidity* associated with earth, the most inert state. Matter associated with the principle of earth has no directional freedom, in the sense that it doesn't flow, so to speak. It involves various degrees of crystalline structures, with relatively rigid and fixed alignment of parts.

The mahabhuta of earth represents increased limitation over a liquid form—such as water into ice when the temperature and motion associated with heat or fire is reduced into a less dynamic state. The mahabhuta of earth is describe as the endpoint of the process of manifestation.

With respect to its correspondence with the fundamental physical forces, it would seem that the mahabhuta of earth is most associated with magnetism—although tangibly expressing all the fundamental particle-forces, and all the other mahabhutas. Is there a basis to associate the mahabhuta of earth more with magnetism and the mahabhuta of water more with electricity?

An electric current flows across objects between charge sources. Attraction and repulsion between two charges occurs in a straight line between the two point sources of the charges. Electric currents generate magnetic fields.

In contrast to the electric force, the magnetic force is a dipole system in which the opposites of attraction and repulsion (north and south poles) are contained in one source in a defined circular path that curves back onto itself—the endpoint of the magnetic field is itself. The magnetic force most tangibly turns back on itself in a closed circular loop around an electric current, in a perpendicular direction to the current flow. This more contained quality of the magnetic force can be thought of as a further limitation compared to the more open flow of the electric force.

All physical matter exhibits magnetic properties in the presence of a magnetic field, and can be classified in terms of degrees to which it is attracted or repulsed by it, depending on the alignment of atoms. In some cases the attractive and repulsive forces cancel each other, resulting in net neutral magnetic properties. The association of the earth mahabhuta with magnetism doesn't mean that all materials made of earth are magnets—though they all interact to some degree with magnetic fields. It is that the abstract principle associated with mahabhuta of earth can be related to the underlying laws of nature that are expressed as magnetism a bit more closely than with the other mahabhutas including water.

The mahabhuta of earth expresses all five mahabhutas, all differentially functioning at this level of nature. The magnetic force is based in the electric charge—consistent with the theory of electromagnetism. Magnetism and electricity are expressed within the limitations of the other three forces (weak and strong nuclear, and gravitational). All constituents are in each other, to different degrees.

Nonlocal mind subtler than the Planck scale? Whether the Planck scale relates to the transition between the gross and subtle levels of nature or it relates to the transition between the subtle and the unified field obviously is a key issue. If it is the transition between subtle and unified field levels, all subtle levels would be within the range of and subject to the limitations of the ordinary spacetime gravitational field. This field is local, quantized in Planck units, limited to light-speed, and subject to the properties of ordinary gravity and thermodynamics. This would mean that all subtle phenomena, including levels of mind, also would be subject to these limitations. This does not seem consistent with progress beyond the physical into the nonlocal level of nature now being proposed as the underlying background to the gross conventional spacetime gravitational field.

Further, it would mean that all subtle phenomena in nature including mind would have to be at a more expressed or larger time and distance scale than the Planck scale, such as at 10^{-32} cm, or 10^{-27}. It is a model of nature with two basic levels: the gross or explicate order associated with the conventional spacetime gravitational field and the unified field, rather than the emerging three-level models which add the subtle level in-between (implicate order).

As discussed earlier, the Sankhya model as interpreted here identifies three fundamental levels—gross relative, subtle relative, and transcendent unified field. Recognizing these three levels of nature—and especially the intermediate subtle relative level now emerging in cutting edge quantum and quantum gravity theories—provides the needed bridge to account for many paradoxes in the reductive physicalist paradigm as well as in the quantum paradigm. Hopefully these speculative matches between fundamental particle-forces and Sankhya will encourage research on further integration of particles and forces in physics.

According to the ancient Vedic model, the 8-fold structure of nature introduced in Chapter 6 and described in more detail in this chapter is reflected in the levels of the subtle inner dimension of mind, and also is reflected in the 8 major organs systems in the gross physical body. [44] The 8-fold pattern is repeated at both subtle and gross levels of the structure of human mind and body.

The most expressed or grossest levels of nature are phenomenally inert, such as atomic particles, elements, chemicals, rocks and earth. The physical universe is a vast expanse of relativistic spacetime containing billions and billions of galaxies. In this vast expanse stars and planets form and decay across eons of time. Recent cosmological theories estimate that the physical universe is about 13.7 billion years old. In the holistic Vedic account consistent with other ancient traditions, however, this would be one aspect of much longer cycles of creation and dissolution.

The material covered so far in this book provides a rational, coherent framework within modern science for a real mind that could interact with real matter. To accomplish this, we needed to establish the concept of nonlocal quantum mind. But the notion of quantum mind does not mean that the mind is quantized in Planck units. Quantum mind is perhaps better understood as relating to a more abstract meaning of the quantum principle, associated with the discriminative intellect and the concept of intelligence in the more abstract sense of structuring form in nature. [52]

A few more steps will be taken now in order to clarify further some of the major dilemmas in quantum theory that directly relate to the apparent fundamental duality of the physical and the conscious mind of an observer. Then we can examine in more detail how mind and matter actually could link together.

Chapter 7

Beyond Mind-Matter Duality

In the Sankhya model of levels of nature, the grosser levels emerge from the subtler levels in a sequence that extends from the completely unified field to subtle mental levels and further to gross physical levels. Again, the levels in Sankhya, including levels of the inner dimension or subjective mind, are cosmic levels extending throughout nature, but also within each of the levels are their counterparts in individual human beings.

The subject/object, mind/matter duality can be said to emerge phenomenally at the level of the intellect (ahamkara). At this level the distinction emerges between manas as subjective observer and tanmatras as subtle objects of sense, which further condense into ordinary physical objects of sense (mahabhutas). It is at the subtle level that the indriyas (organs of sense and organs of action) also emerge. Thus we have the observer as mind, the senses, and the objects of sense. The indriyas (including senses) involve perceptual and behavioral interactions with subtle and gross objects. Details of these levels will be discussed in Chapter 8 on levels of mind.

The processes of phenomenal manifestation of these levels can be likened to sequential symmetry breaking of the unified field. But in this view the unified field is consciousness itself. This will be discussed further in Chapter 8, and also in Chapter 10 on epistemology and the means of gaining knowledge.

To add more clarity to the notion of nonlocal mind and how it goes beyond the physical, it may be useful to consider again contemporary views in quantum physics of the role of subjective mind as it relates to objective matter, and to take the discussion a few steps further. We will first examine the model in orthodox quantum theory of the collapse of the quantum wave function via an observer, which bred the measurement problem that has been so paradoxical in quantum theory. A somewhat different angle from Chapter 2 will be taken here.

Realism vs. idealism. The models of quantum wave function collapse outlined in Part I could be understood as attempts to address the long-standing philosophical debate of *realism* versus *idealism*. The realist assumption of an objective world—a mind-independent world existing in-itself apart from the observer—is fundamental to classical science. This view further assumes that for the most part the objective world can be measured without the observer influencing it. As Einstein [19] asserted:

> "An essential aspect…of things in physics is that they lay claim, at a certain time, to an existence independent of one another, provided these "objects" are situated in different parts of space…. Unless one makes this kind of assumption about the independence of the existence (the "being-thus") of objects which are far apart from one another in space—which stems in the first place from everyday thinking—physical thinking in the familiar sense would not be possible. It is also hard to see any way of formulating and testing the laws of physics unless one makes a clear distinction of this kind." (p. 154)

In contrast is the idealist perspective, stated succinctly in the following quote of cognitive neuroscientist Francisco Varela, neurophilosopher Evan Thompson, and cognitive psychologist Eleanor Rosch: [53]

> "The idealist, on the other hand, quickly points out that we have no access to such an independent world except through our representations…. In fact, we simply have no idea of what the outside world is except that it is the presumed object of our representations. Taking this point to the extreme, the idealist argues that the very idea of a world independent of representations is itself only another of our representations…." (p. 161)

Revisiting quantum wave function collapse. Taking the realist perspective, modern science progressed with great success applying the objective experimental method at *macroscopic* scales. But as *microscopic* scales have been investigated, the viability of the realist assumption of the independence of observer and observed has been severely challenged—a consequence of quantum theory and the model of wave function collapse. According to the initial orthodox interpretation, an elementary matter particle is no particular *thing* prior to being observed or measured. It is characterized mathematically as a *probability wave function*. Only with an observation does the quantum wave function translate into a classical physical object with definite measurable dynamic attributes in ordinary space and time.

Though particles are represented mathematically in classical physics as points in space with *no extension*, in quantum field theory quantum waves are envisioned as having *infinite extension*. Superposed abstract quantum wave functions are theorized to collapse into discrete particle-like objects upon observation. But this glosses over the distinction between the quantum wave function as a concept in imaginary mathematical 'space' and classical objects in *real* physical space.

The quantum wave function can be said to be a mathematical *reality* (associated with deductive reasoning), and states of the object are a sensory *reality* (associated with empirical experience and inductive reasoning). The conflation of these levels has led to the measurement problem and the model of instantaneous collapse of the quantum wave function. Resolution of these issues requires unpacking into an expanded ontology, now emerging in evidenced-based scientific models and described in detail in the Sankhya model.

In the following quote, string theorist Brian Greene [10] expresses concern about the model of quantum wave function collapse as proposed in the initial orthodox (Copenhagen) interpretation:

"According to standard quantum mechanics, when we perform a measurement and find a particle to be here, we cause its probability wave to change: the previous range of potential outcomes is reduced to the one actual result our measurement finds.... In the standard approach, the collapse happens instantaneously across the whole universe.... Nevertheless, there is a hitch. *After more than seven decades, no one understands how or even whether the collapse of a probability wave really happens."* (p. 19)

With progress beyond the orthodox interpretation, there has been an unpacking of the processes involved in getting to a real physical object and the role of the conscious observer. Quantum waves have gained ontological status as real objects in nature, independent of the conscious observer. This important step of progress can be applied to clarify wave function collapse using a different perspective that does not *objectify* this theorized process. In this view the 'collapse' refers to a change in the subjective knowledge state of the observer, [51] not a change in the real object observed.

Before making an observation, what is *real* about the state of the world *from the observer's perspective* is probabilistic rather than discrete and definite. After observing it, what is *real* about the state of the world *from the observer's new perspective* is that it is in a discrete physical state. The observer's state of knowledge changes or 'collapses' upon observation—not the objects sensed in the world.

The 'collapse' can be viewed as simply the natural updating of knowledge in the observer, occurring whenever an observation is made. But whether observed or not, the physical world is in a *real* state independent of any observation of it. In the following quote, physicist Christopher Fuchs [54] anticipates this view:

> "When a quantum state collapses, it's not because anything is happening physically, it's simply because this little piece of the world called a person has come across some knowledge, and he updates his knowledge... So the quantum state that's being changed is just the person's knowledge of the world, it's not something existent in the world in and of itself." (p. 42)

This view of the 'collapse' as pertaining to the knowledge state of the observer and not the observed is consistent with ontologically real quantum processes underneath the 'inviolable wall.' It is progressing toward a logical understanding of how the change takes place in nature from quantum *possibilities* to physical *actualities* independent of an individual conscious observer.

From another angle, in orthodox quantum theory the wave function is an equation in imaginary mathematical space that instantaneously collapses into a discrete classical object when measured or observed by a conscious observer. But the observer, the measuring device, and the object it collapses into are all classical, real physical objects. These classical physical objects don't have an imaginary quantum wave function in them. The measuring device itself doesn't have conscious awareness at least in the sense of what is ordinarily meant by consciousness, and doesn't have a mathematical wave function in it either.

When an idealized mathematical model is imposed on wave and particle properties of objects, it compels the conclusion that wave function collapse is needed to get from the mathematical model to ordinary objects. The inner subjective updating of the knowledge associated with wave function 'collapse' has been *objectified* onto the objective world, leading to paradoxes in the orthodox interpretation.

Physical objects are experienced as discrete independent objects by virtue of how the ordinary senses function. When the inner cognitive functions of intellectual imagination impose a mathematical wave model onto nature, it seems to require a collapse of the wave function in order to obtain concrete classical objects.

The 'collapse' can be said to be on the level of mathematical conceptions in the mind—as a reduction of possibilities and probabilities into specific states of knowledge. There is a 'collapse' of the state of knowledge from mathematical possibilities to one definite outcome based on new information from making an observation. This doesn't mean a collapse of a quantum object into a physical object.

An overriding issue has been emerging that contrasts the view of the cosmos in terms of locality—captured in the concepts of local causality and the light cone in relativity theory—with entangled nonlocality associated with the concept of *holism* in quantum theory. Bohr's undivided wholeness of the experimental arrangement, von Neumann's quantum wholeness or all-quantum *reality*, the principle of nonlocality, and Bohm's psi wave and implicate order represent steps toward a more interconnected, nonlocal model of the universe.

Historically modern scientific methodology hinged on the assumption of the independence of objects in nature (refer to Einstein's quote earlier). But as Greene [10] points out, this picture fundamentally changes with nonlocality:

"We used to think that a basic property of space is that it separates and distinguishes one object from another. But we now see that quantum mechanics radically challenges this view. Two things can be separated by an enormous amount of space and yet not have a fully independent existence... Space, even a huge amount of space, does not weaken their quantum mechanical interdependence." (p. 122)

Cutting edge models in quantum physics suggest that the object independence on the surface of nature is underlain by increasing interdependence at deeper levels. And from the holistic perspective, all things are connected and no thing is completely independent. It implies that the independence of objects, objective reality, and subject-object duality are conditional or contextual views of nature, with limited ranges of application.

Herein is a major concern relevant to orthodox as well as non-orthodox quantum theory, and also relativity theory. The model of quantum wave collapse throughout the unbounded quantum field implies nonlocality and holism, not locality and reductionism.

In addressing the measurement problem, quantum wave collapse, and the mind-body problem, contemporary models posit not two but rather three ontological levels. Importantly the expanded three-level ontology points to a resolution to the fundamental issues of locality and nonlocality, reductivism and holism, and the mind-body problem. It also leads to resolution of unresolved paradoxes associated with quantum wave function collapse.

To summarize much of the discussion so far, and to unpack further quantum wave function collapse, key issues will be laid out in the form of questions and brief answers that incorporate the more expanded and integrative Sankhya perspective. Many of the issues considered in the book so far are addressed in this short summary of points to unpack the model of quantum wave function collapse.

Do objects in nature exist independent of the process of observing and the observer? In the Sankhya model the duality of subject and object fundamental to classical science emerges at the subtle nonlocal level of cosmic intellect. At the subtle level of nature, there is a separation into universal mind including individual minds (manas) with sensory ability to experience objects (indriyas) and the subtle objects of experience (tanmatras).

This subject-object separation, commonly associated with independence of objects and with particle interaction causality (mahabhutas), is much more predominant on the ordinary gross physical level of nature. Thus there are objects in nature that can be said to be independent of the process of observing and any individual observer on both subtle and gross levels of nature, which supports realism and fundamental objectivity.

On the other hand, idealism and fundamental subjectivity are also supported in that what the independent world appears to be depends on the subjective observers. Both of these *real* gross and subtle levels are conditional because ultimately all objects and observers, all objectivity and subjectivity, remain within the infinitely self-interacting unified field or universal Being (Prakriti/Purusha). Objects do exist relatively independent of individual subjective minds in the phenomenal universe, but not independent of the unified field.

Is there a level of quantum reality in nature? Quantum *reality* would mean that quantum objects are ontologically *real* objects in nature, and Sankhya supports this view as a conditional *reality*. Quantum objects (but not in Planck units) are not physical in the sense of being composed of classical particles localized within ordinary relativistic spacetime within light-speed. They can be viewed more in terms of *real* nonlocal objects, for the most part associated with the tanmatras, the subtle objects of nature (again, but not Planck unit quantized).

Is there a nonlocal field not mediated by the four known fundamental local forces? Yes, this ontologically real subtle level, medium, or field of nature does not have the limitations that characterize the four known local forces (gravitational, strong and weak nuclear, electromagnetic), at least as ordinarily understood. Nonlocal processes in the subtle level are associated with the tanmatras, which can be said to be entangled and interdependent. Going from subtler to grosser levels, the more interdependent wave nature of objects is suppressed or hidden and the more discrete, independent Planck unit quantized and particle nature of objects becomes more prominent. In other words, the reductive parts of nature, rather than the underlying wholeness, become more apparent from transcendent to subtle to gross levels.

Is superluminal communication possible? As depicted in Figure 1 in Chapter 5, the subtle nonlocal level includes the subtle objects of sense (tanmatras), the subtle senses (gyanendriyas, and also karmendriyas), and the levels of mind (manas, ahamkara, mahat). Processes at this level are entangled and their interactions are not limited to ordinary space and time and the speed of light.

In this view superluminal communication is how mind naturally influences matter. At the gross local level of nature, change is via particle interaction mechanics within the speed of light. At the subtle nonlocal level, it is via nonlocal causal wave dynamics that can be superluminal but are not instantaneous. At the level of the infinitely self-interacting unified field, it can be said to be instantaneous—that is, beyond both the ordinary gross independent and subtle interdependent levels of spacetime—*infinitely self-interacting,* or *Self-interacting.*

Are conscious intentions and free will causally real? Yes. In Sankhya the subtle domain (or implicate order) is where individual conscious minds exist as nonlocal phenomena. Individual minds are underneath and permeate local physical body/brain processes; and conscious intentions are subtle impulses that drive the subtle and gross organs of action in the body and brain as a causal wave according to the free choices of the individual observer.

In other words, individual subjective minds impel changes in phenomenally *real* inert matter via subtle thought waves of *real* mental intentions—mind over matter. It is important again to point out that nonlocal mind does not mean that it cannot be identified as individual mind. It does not mean some sort of group mind with no individuality to it.

Is nature fundamentally indeterminate and random or deterministic even if unfathomable? Nature is determinate, not random and indeterminate (though random processes certainly seem to be involved on some levels). Causation does not end at the 'inviolable' quantum level such that quantum indeterminism takes over entirely.

However, in this view causation is not just the discrete 'billiard ball' or 'particle interaction' models, but a nonlocal causal wave. At the subtle level, causation is more wave-like than particle-like, expanded into a more abstract, wider-angle, interdependent medium or domain of space, time, and causality. It can be envisioned as somewhat analogous to the notion of spacetime dilation in relativity theory, but not with respect to ordinary spacetime. The determinism is so complex that it is probabilistic with respect to calculations.

Is there is a collapse of the quantum wave function? There is no quantum wave collapse in the orthodox sense that it occurs instantaneously by a conscious observation. Also, there is no quantum wave collapse in the sense of objective reduction due to decoherent interaction with random influences in the physical environment.

However, the probabilistic knowledge based on the mathematical quantum model of probabilities can be said to 'collapse' into classical discrete non-probabilistic knowledge when the object is observed. The closest thing to a 'collapse' is the updating from probabilistic to definite knowledge based on sensory observation of the consensually-agreed upon *real* world of individual observers. But then how does Nature as a whole, the unified field of Being, change from possibilities to physical actualities apart from individual observers?

In the Sankhya model the infinitely self-interacting unified field of universal Being manifests into finite actualities of individual observers and individual objects. This process occurs naturally in a manner similar to spontaneous sequential symmetry-breaking. It is in this broader context that decoherence can be understood, not as individual objects 'decohering' via interaction with the environment but rather as levels of nature expressed in increasing entropy.

This contrasts with the decoherence model in which individual objects collapse from quantum waves to classical objects. In this Vedic model the gross physical environment of particle interactions manifests from the subtle level at almost the same time; and then the particles combine to structure microscopic and macroscopic objects in the gross physical level through ordinary time. It is on the subtle level where the subject-object duality becomes expressed, prior to the gross level, covered in detail in the structure of the Veda itself.

When an observation is made by a real conscious observer of the real and already existing subtle and gross natural world, the observer gains specific knowledge of it. This reflects the self-correcting scientific method of empirical validation fundamental to science. Validation through empirical experience updates and advances theoretical knowledge of it. Thus the model of quantum wave collapse via decoherence is subsumed in processes described in terms of sequential symmetry-breaking. In this view the notion of 'collapse' applies to a change in knowledge state of the observer from mathematical possibilities to specific knowledge upon observation.

However, the term 'collapse' also has been applied to the infinitely self-interacting, self-referral dynamics of the unified field of consciousness.[3] It is used to describe intellectually the transcendent unified field as containing both whole and part, universal and individual, at the same time so to speak, one collapsing into the other infinitely and instantaneously. In this completely abstract sense, infinity collapses to a point as the self-interacting dynamics of the unified field. But also, the point is a singularity that is the whole, and It collapses into infinity. Each point is infinity. In other terms, infinite silence collapses into infinite dynamism, and infinite dynamism collapses into infinite silence. This begins the discrimination of phenomenal parts that ultimately remain the wholeness. It refers to the dynamics of manifestation that is the structure of the Veda.

Every phenomenal thing remains in the unified field and is ultimately nothing other than the infinite eternal silence, but appears to manifest into discrete forms in our familiar natural world through the self-interaction of part and whole, point and infinity, dynamism and silence. As Maharishi[55] recently stated:

"Silence and dynamism—they are one thing, not two things."

Hopefully this overview provides a framework within the context of modern scientific theories that go beyond reductive physicalism in order to proceed with the overall discussion about how nonlocal mind and local matter link together. We will now consider in progressive detail this issue, in general accordance with the Sankhya perspective.

Gross physical and subtle non-physical

In Sankhya the *grossest* of the subtle tanmatras—the essence of the earth constituent—is further restricted by increasing tamas into the *subtlest* of the gross relative level—the *mahabhuta* of space. The space mahabhuta contains all five qualities of the subtle essence tanmatra of earth, but the four other qualities are not as expressed.

The change from the subtle essences of the tanmatras to the gross constituents of the mahabhutas might be said to involve predominantly nonlocal wave dynamics condensing into local point-particle mechanics, but at this stage the wave notion is more general. The nonlocal properties of the tanmatra of earth—with the five essences on the subtle level corresponding to the five senses—condense into local relativistic spacetime, the mahabhuta of space.

It again may be useful to discuss subtle versus gross wave processes. A wave is a progressive vibration through a medium, such as air or water, without corresponding progress of the parts or particles. The medium undulates in opposite directions, curving back onto itself through points in time as impulses of propagating energy. This can be thought of as a pulsation that phenomenally gets extended into a wave as it moves through space and time—expanding and contracting in all dimensions simultaneously. An ordinary physical wave of water as a pattern of the movement of particles, such as an ocean wave, is a collection of undulating localized water molecules. However, a subtle wave is nonlocal, much more spread out, and not a collection of particles or molecules.

The five mahabhutas can be thought of as point-particle fields with progressive limitations, each more expressed one embedded in the previous one and adding an additional quality. In terms of the gross senses, they are the ordinary qualities or constituents associated with space, air, fire, water, and earth.

From the mahabhuta of gross space emerges the four other mahabhutas—the *paramanus* or smallest indivisible units. The progressive unfoldment of the five mahabhutas involves increasing restriction of motion, localization, symmetry breaking, and phenomenal independence—denser packing and rigidity, so to speak, associated with increasing tamas akin to the concept of mass.

Within the framework of locality, one way to think about the paramanus is that they are structured by the relativistic spacetime gravitational field being further limited, drawing into its point value—collecting into or curving back onto itself. In sequential manifestation from transcendental to subtle to gross, the field can be thought of as curving back onto itself or as compactified into discrete forms that function as phenomenally independent and self-contained objects.

The mahabhutas are sometimes described as dimensionless points. As paramanus, they exhibit the properties of discrete matter particles, gaining mass and extension in space, which also characterize Planck-unit quanta in quantum theory. These mechanics of manifestation in the gross relative domain may have their counterparts in the contemporary models exemplified by M-theory, string theory, and loop quantum gravity theory.

These theories can be viewed as attempts to characterize how nature structures Planck-unit quanta and fundamental particles from dimensionless points. The theories seem to be attempts from a reductive perspective to explain how spatial extension in conventional local spacetime is manifested in the transition from subtle to gross creation—from predominantly nonlocal wave-field dynamics to point particle, Planck-unit quantum mechanics, and particle interactions.

The subtle nonlocal field or level would not have the limitation of light-speed, physical mass, Planck-unit quanta, or be subject to the gravitational force associated with conventional gross spacetime. This doesn't mean that they are not relative, not finite, not deterministic, and not subject to an attractive force or attractive quality; but rather that they aren't as limited as objects built of gross matterstuff.

Again, we can think of three levels of curving back or referral processes. At the transcendental level of the unified field the referral process is *infinite* self-referral, or *Self*-referral. At the subtle relative or subjective level the referral process can be characterized more in the sense of referral to the individual self—finite individual self-referral moving toward infinite self-referral as individuality is established in universality. At the gross relative or objective level, the referral process can be characterized as point-particle referral—referral into a point value, more commonly called *object-referral*.

In the increasing limitation, the 'curving back' eventually curves back onto the local point value at the gross level of existence, and further to the Planck-unit quantum and then atomic structures. These are highly localized forms that appear to be cut off from the underlying nonlocal and infinite levels in which they are embedded. The subtle levels—as well as the transcendent level—become *unexpressed* or *hidden* at the gross, discrete, highly localized level—via the hiding influence of tamas.

The gross level expresses *least* the inherent qualities of life and intelligence. It can be characterized as being objectified and inert—not exhibiting subjectivity or life. Gross objects take on the appearance of being independent of each other—exhibiting localized interactions in classical physics. They have a localized particle aspect, but also a more hidden but fundamental nonlocal wave aspect.

Gross objects made of particles appear as if they don't have the ability to initiate action or movement—inert matter, with no power of agency. They are said to be inherently dynamic, but don't appear to exhibit agency, intentionality, or intelligence 'on their own.' With respect to living organisms made of phenomenally inert particles, it is the influence of the energy and intelligence on the subtle relative level that guides their orderly, intelligent, negentropic behavior.

Each of the five gross elements (mahabhutas) exhibits a specific quality that corresponds directly to one of the five sensory capacities. As the limitations of the five elements increase, their gross objective forms express more specificity, which correspond to tangible sensory qualities. Whatever additional limitation characterizes the next grossest level of nature within the gross relative domain, it brings out its corresponding sensory quality. Space relates to hearing, air adds touch; fire adds sight; water adds taste, and earth adds smell. The grossest mahabhuta of earth expresses all five qualities, and thus can be sensed via all five sensory modalities.

The mahabhutas are progressively associated with increasing tamas, increasing limitation, increasing entropy, increasing sensory specificity, and with decreasing degrees of freedom and symmetry. On the other hand, they are ultimately the unified field appearing to interact as phenomenal nature while remaining within itself.

So in a brief picture, how does all this tie together? The unified field or universal Being is the infinite eternal, perfectly orderly source of all intelligence and energy. It manifests spontaneously in sequence from the subtlest levels that are the most encompassing and powerful finite levels to the grossest and more limited finite levels, each the basis for the next level while remaining within the unified field.

From the unified field level that can be characterized as the unity of point and infinity, instantaneity and eternity, phenomenal manifestation and dissolution occurs across vast eons of cycles akin to sequential symmetry breaking, in which the infinite value appears to be increasingly hidden and limited and the point value appears to be increasingly prominent, from complete interdependence to the appearance of independent physical objects in ordinary space and time. This can be conceptualized in terms of unbounded wave dynamics becoming restricted to bounded particle mechanics.

The transition from nonlocal causal wave dynamics to local particle causal mechanics is associated with the change from subtle to gross—implicate to explicate orders. It is not that nonlocal wave dynamics are no longer relevant at gross levels of the explicate order, because the particles are embedded in them. Particle qualities appear to predominate on the surface, and wave qualities at deeper levels.

The three forces of attraction (sattva), dynamic action (rajas), and resistance to change (tamas) move toward more restriction (increasing tamas) from subtle space to Planck-unit quantized space. This defines the Planck scale, related to light-speed as the textural limitation of the quantized spacetime gravitational field. The wave pulsations of subtle space condense further into particle mechanics in the relativistic spacetime gravitational field that are more discrete, localized, and tangible from the perspective of sensory processes restricted to the same mechanics. The spacetime gravitational field (related to the mahabhuta of space) further condenses through spontaneous sequential symmetry breaking into the strong nuclear, weak nuclear, and electromagnetic force fields (associated with the other mahabhutas of air, fire, water, and earth). The vast diversity of ordinary visible objects in the gross physical universe is formed from these constituent quantum fields.

The process of cosmic evolution continues, through which these constituents that appear on the physical level to be inert—because their inherent dynamic intelligence is hidden and latent—congeal into stars, galaxies, and planets. Eventually from the apparent latent order inherent in these objects emerges living beings that are complex enough to express higher and higher levels of dynamic intelligence toward the ultimate totality of the unified field.

Again, the transition from subtle to gross relates to the change from the tanmatra of earth to the mahabhuta of space. The tanmatra of earth is the grossest of the five subtle tanmatras, and the mahabhuta of space is the subtlest of the five mahabhutas. Quantum gravity theories seem to be attempts within quantum physics to begin to address this transition from subtle to gross, though these theories generally do not have a holistic framework of understanding to characterize them as does Sankhya, but rather a reductive framework.

To relate in more detail the Sankhya view to modern physics, the gross mahabhuta of space seems most akin to the notion of the relativistic quantized spacetime gravitational field, associated with quantization such as Plank-unit wave packets, strings, branes, and particles. The subtle tanmatra of earth seems to be what is attempting to be envisioned with concepts such as for example information space in loop quantum gravity theory and zero-branes in M-theory.

In the next chapter, the Vedic approach to the subtle, subjective levels of mind will be discussed in more detail. The links between the physical and corresponding deeper mental forces or qualities are outlined in a profound holistic model that extends across gross and subtle levels of nature. This ancient Vedic system of levels of nature reflects a deep integration of subjective psychological and objective physical functions. No theory with such a high degree of integration has yet been envisioned in modern science.

Chapter 8

Levels of Mind

Beginning to formulate mathematical models of the ultimate unity of nature in unified field theory is a great accomplishment in modern science. It also has established the theoretical basis to link up with the ancient Vedic knowledge system. Only in recent decades has modern science glimpsed deeply enough into nature to link with that view, as well as other ancient traditions that posit ultimate unity.

Previously thought to contrast with modern scientific accounts, ancient Vedic views are now being corroborated by contemporary formulations that provide similar descriptions of an infinitely dynamic, self-interacting unified field. [52] Until modern science arrived at a rational theoretical framework for developing a unified field theory, the correspondence with ancient holistic views generally was not recognized. The most parsimonious explanation for the correspondence is that these knowledge traditions converge on the same unified field from their respective vantage points. [52] Logically there is one completely unified field; and logically it unifies all of nature, including objective and subjective.

The unified field as consciousness

In the ancient Vedic tradition, as well as other traditions albeit with various cultural and language differences, there is purported to be a transcendent universal essence of nature. This is also a core principle in the 'perennial philosophy,' which attempted to summarize and integrate ancient and modern views. [56] Most of these ancient traditions further hold that direct experience of the transcendent universal essence of nature is possible. It is said to be possible because the universal essence of nature also is held to be the essence of one's own consciousness—universal *Being* as the essence of individual being, universal consciousness as the essence of individual consciousness. In the holistic Vedic account the unified field is universal Being, consciousness itself.

That view contrasts dramatically with the physicalist view of consciousness as reducible to neural activity in the local physical brain. As described earlier, the physicalist account can be characterized as a matter-mind-consciousness ontology, in which conscious mind somehow emerges in and is entirely dependent on unconscious processes in the physical brain. The Vedic account can be interpreted in the opposite way as a *consciousness-mind-matter* ontology, in which the unified field of consciousness or universal Being is the basis of mind and matter. [5] The following discussion brings out a few more points that may help further to understand the view that the unified field as the source of everything also would be the source of consciousness—even consciousness itself.

In unified field theory all change would originate from and take place within the infinite eternal unified field, ultimately caused by the unified field interacting with itself. The unified field as the source of everything would be the *infinitely self-interacting* cause of all phenomena. Anything that exists and interacts with other things would be ultimately the unified field interacting with itself.

A self-interacting field is modeled by non-linear topologies that reflect back on themselves and create new structures. The ultimate unified field, as an infinitely self-interacting field, would be able to create any actuality, to manifest any possibility that could exist. It would be the source of all the laws of nature that govern orderly change. And it would discriminate distinctions within itself that create the vast diversity throughout phenomenal nature.

If the unified field is the source of all order in the universe, and that order emerges somehow only within it and due to its self-interacting dynamics, it could be said to exhibit attributes of a lowest entropy field of perfect order. It would include negentropic effects— negative entropy and order-creating processes—typically associated with the concept of intelligence.

It thus seems reasonable that the unified field also logically could be considered a field of intelligence. This is the case at least in the sense of a self-sufficient source of order and the source of all the laws of nature.

Intelligence also connotes *intentionality*—purposive action or agency. Is there a reasonable intuitive sense in which the unified field can be attributed the property of intentionality? Certainly in the sense of the initiation of orderly action the unified field can be thought of as intentional—as the ultimate, singular, uncaused cause and origin of all action in nature—the First Cause.

Another connotation of intelligence is self-knowingness or self-awareness. Is this quality also attributable to the unified field?

This issue can be viewed as related to the distinction between *self-interacting* and *self-referral*. If the unified field is infinitely self-interacting, inherently dynamic, and inherently orderly (generating active intelligence), would it then also be inherently self-referral or referring back to itself, in the sense of being self-aware?

One approach to this issue is to identify the characteristic marks that something is conscious or self-aware. Typical signifiers that an entity is conscious relate at least in part to the signifiers of life such as intelligence, intentionality, attention, and the survival instinct. These can be related to the notion of discriminative action by an entity to maintain its own existence, to survive. Can the unified field maintain itself? That is, is it self-sustaining?

This might be thought of in part as an empirical question of whether the universe has existed and continues to exist in some form or another. It would seem consistent with ordinary experience that the universe does continue to exist in some form. This would be by virtue of being a manifestation of nothing other than the infinite, eternal, self-interacting unified field because nothing else exists. If the signs of its existence—that is, creation and created beings such as us—are maintained through time; and in addition are sustained by its own self-interacting dynamics, then the unified field would seem to be self-sustaining. This is consistent with at least one of the signifiers of self-awareness. Are there other signifiers of the unified field as an ultimate level of nature that is self-aware, conscious of itself?

Some approaches to consciousness assume that an entity has to have a body that supports conscious experience in order to be conscious. In what ways might the unified field perspective satisfy this assumption?

If the unified field is the source of everything that exists, it necessarily would be the source of all bodily forms that support conscious experience. The unified field could be thought of as its own body—the basis of the abstract principle of form, the essence of any particular body, completely self-referral.

Alternatively it could be taken that the body of the unified field is the entire manifest universe as an integrated unit—an embodied, unitary, completely self-sufficient system. This is consistent with increasingly popular ideas that the universe itself is a biological organism, such as the 'living universe.' These points suggest that the unified field can be thought of as having a body that could support consciousness.

However, most approaches to consciousness would add the additional requirement that the 'body' needs to have a high enough degree of complexity and interactivity for conscious experience to emerge. But there would be no degree of complexity and interactivity higher than the total integration of everything in the unified field. Any localized functional or structural aspect at any level of nature would be, in one sense, less complex and less interactive than the totality as an infinitely self-interacting unified field.

The unified field would necessarily contain all substances and functions comprising any of its parts. There would be no greater wholeness, no greater capability, including the capability of self-awareness, than the total capability of the unified field. As the source of everything, it obviously has a deep correspondence with the concepts of both the personal God and the impersonal Godhead as infinite and eternal that are found in many belief systems.

The unified field certainly would be complex enough to support consciousness—infinitely complex as including everything, and infinitely simple as the singular totality of everything. As the basis of subjectivity and objectivity, it would contain or even be consciousness. Thus there are indirect empirical, logical, and intuitive bases for the unified field as being self-referral, conscious of itself.

These points on the relationship of unified field theory and consciousness relate to long-standing philosophical questions. Here the intent is to provide a glimpse at how ancient traditions of knowledge address these deep questions rationally. It also suggests how contemporary unified field theory can be understood to be consistent with the holistic view in these ancient traditions.

In the perspective of Sankhya the ultimate, infinite, eternal, indivisible wholeness of the unified field phenomenally manifests into differentiated parts. The parts reflect lesser and lesser degrees of wholeness until the wholeness becomes as if completely hidden, latent, and inaccessible. At this level the whole is in terms of an independent unit or object. From universal Being emerges the appearance of inert, unconscious, lifeless atoms and elements. These lifeless phenomenal parts are said to be the building blocks of nature in reductive physical theory.

In Sankhya the levels of mind can be understood to be phenomenally real levels of nature in the underlying transcendent unified field of consciousness itself. Although ultimately all levels of nature are nothing other than the infinite eternal unified field of consciousness itself, in the enumeration of levels of nature in Sankhya the levels of mind can be thought of as not conscious. They are thought of as natural processes that present sensory phenomena to consciousness itself. In this sense, both mind and matter can be viewed as not conscious in themselves.

This view in Sankhya may seem similar to reductive eliminativist views in cognitive science consistent with physical theory. These views suggest mind and consciousness are epiphenomenal, don't really exist, and are completely accounted for by physical brain processes. But in Sankhya the view that mind and matter are not conscious is an important stage of understanding to distinguish mental activity from pure consciousness, associated with the concept of *Maya*. It is important also to place this view in the bigger picture of the Vedanta perspective that all levels of nature are nothing other than phenomenal fluctuations of the ultimate universal Self. In that sense, all levels of nature are nothing other than consciousness.

To examine further the relationship between mind and matter, a crucial link is the subtle, non-physical level. This is now being glimpsed and speculated about in quantum physics, but not yet adequately articulated.

To understand the relationship between mind and matter in Sankhya, the difference needs to be clarified between the *gross* mahabhutas and the subtle *tanmatras* (See chart in Chapter 5). Again, in Sankhya the mahabhutas refer to constituents that compose the gross physical domain, which has been the primary focus of the extensive rigorous experimental research in modern science. All the time and distance scales identified in modern science are said to comprise the gross relative domain. This includes galactic clusters, galaxies, solar systems, stars, planets, nations, states, communities, families, individuals, organ systems, cells, molecules, organic and inorganic compounds, elements, atoms, elementary particles, and possibly strings, branes, and quantum fields if subject to the limitations of the relativistic spacetime gravitational field.

Although there are many degrees and layers in this vast range, they are identified as the gross relative domain. It also includes all ordinary objects subject to ordinary gravity experienced at the macroscopic level via the ordinary senses. Sankhya delineates an additional non-physical domain, now emerging in cutting edge quantum theories. In this subtle domain are levels of mind, and also non-physical objects of sense composed of the five tanmatras.

Thus there are five gross constituents (mahabhutas) associated with particle-like mechanics, and correspondingly five subtle essences (tanmatras) associated more with wave-like dynamics. In the interpretation of Sankhya discussed here, these subtle objects are not made of particles or Planck-unit quanta, and not subject to limitations of light-speed, the particle interaction model of causality, and the relativistic spacetime gravitational field. From the perspective of the ordinary senses, the subtle essence constituents are intangible and rarely experienced. But in this holistic view, in more advanced experiences they are comparatively more *real*. They are closer to and more reflective of the infinite power of the unified field. They are experienced as more energetic, inherently dynamic—self-luminous so to speak—embodying more the essence of intelligence and order.

The five gross constituents and the five subtle essences of the objects of sense also correspond to gross and subtle functions of the five senses, called the *gyanendriyas*. In this view the subtle level embodies much more fully the qualitative sensory essences of the five elements or constituents in more refined sensory experience. [5] Again, according to Sankhya the gross objects of sense also can be said to have existence on the subtle level. However, this is not necessarily the case for subtle objects. Many objects in the subtle relative domain are not manifest in the gross relative domain—indeed even most of them seem not to be. These phenomenal sensory objects are made of the subtle tanmatras, but not also the gross mahabhutas.

In the Sankhya perspective the gross material domain can be described as the crusty surface of a vastly more expansive universe. This more expansive universe is said to have many layers of subtle creation and subtle objects of sense composed of the subtle essence elements (tanmatras), as well as even subtler mental levels.

As suggested by the chart in Chapter 5 (Figure 1), because the five senses are subtler than both the gross elements (mahabhutas) and the subtle essence elements (tanmatras), they have the potential to sense both of these levels. This meaning of the senses doesn't refer merely to the ears, skin, eyes, tongue, and nose and corresponding mechanisms in the brain, but rather a much more abstract meaning.

In the ordinary waking state, however, sensory experience is almost exclusively limited to experience of macroscopic objects in the gross relative domain limited by ordinary gross functioning of the senses. In this range of perception, subtle experiences are hidden due to undeveloped sensory processes. If experienced at all, they are held to be highly significant but elusive, intangible, rare, or even believed to be illusory and sometimes questioned as dysfunctional and not distinguished from illusory hallucinatory experiences in altered waking state consciousness or in the dreaming state.

It is not, however, that experience of subtle levels means that humans necessarily can see directly the finer-grained time and distance scales in ordinary space and time such as atomic, quantum, or Planck-scale structures. In the same way that the gross functioning of the ordinary senses relates to the levels of ordinary macroscopic objects underlain by unseen particle-like properties and mechanics, the subtle levels of sensory perception relate to subtle objects underlain by unseen wave-like dynamics.

Generally phenomenal experience involves the results of processing as objects of sensory perception, rather than the underlying mechanics or dynamics themselves. This seems to be the case in both ordinary gross and extraordinary subtle sensory levels. Sensory experience is typically of a wholeness or gestalt of specific objects, not the underpinning mechanics or dynamics of the objects.

Focus on the objective gross relative level of nature characteristic of the physicalist worldview is certainly reasonable given the belief and experience that it is the only domain of existence. In the physicalist paradigm the intelligent negentropic activity of living organisms is observed on macroscopic and larger microscopic levels, but not directly at tiny time and distance scales. At even smaller scales, only non-living physical energy/matter is believed to exist.

Given this view, it is also quite reasonable to assume that life and conscious intelligent behavior emerge only at larger time and distance scales. However, a coherent account of the emergence of life, mind, and consciousness from non-living matter and energy has not been achieved. This is also quite reasonable from the holistic Vedic perspective, because in this view physical theory is an incomplete account of the ontological structure of nature.

Importantly Sankhya provides an explanation for the phenomenon that material objects at smaller microscopic and ultramicroscopic scales don't 'on their own' exhibit life, intelligence, and conscious intentions. These objects are composed of gross elements in which these capabilities are hidden and unexpressed (mahabhutas). The total potential of nature exists in them, but in latent form.

Though well-designed computers, cellular automata, or Turing machines made of gross physical matter may mimic such 'intelligent' behaviors effectively in the gross domain, they are not sentient beings from the Sankhya perspective. Macroscopic physical systems exhibit conscious intentional behaviors as sentient beings only when they are part of living organisms with functioning physiologies.

In the mainstream physicalist view, when the individual physical body stops functioning the existence of the individual organism as a sentient being ceases. Most cultural and knowledge traditions, however, hold a different view. In this alternative view the individual sentient being is not just an objective, gross, physical body but also a non-physical, subjective self or individual 'soul' that occupies a gross physical body. This theorized intangible soul understandably has been difficult to define and verify within the mainstream scientific paradigm fixated on the gross surface level of nature as the only real level of existence.

With advances in modern science outlined in this book, theories of a non-physical domain not entirely accounted for by the gross physical brain/body are necessarily emerging in pursuit of a coherent account of nature. These advances establish a rational framework for understanding subtler levels, which can be understood to be where the soul traditionally has been placed in most ancient knowledge systems, even if not clearly articulated in them with respect to the language of modern science.

In Sankhya the mahabhutas are the gross physical elements, the tanmatras are the subtle essence elements, and the indriyas are the even subtler functional senses. The individual sentient being or soul is composed of these and other subtler subjective levels. All the non-physical levels identified in the chart in Chapter 5 are said to comprise the individual soul (tanmatras, indriyas, manas, ahamkar, Prakriti, and Purusha, but not the gross mahabhutas).

The levels of mind in the Sankhya chart will now be shown to be comparable to functional levels that have been developing in various theories over the past 150 years of progress in scientific psychology, cognitive science, and neuroscience. But these psychological models had not been integrated into a coherent functional theory of mind. This progress will now be briefly summarized.

Levels of the inner dimension

In behavioral psychology the gap between stimulus (S) and response (R) was a 'black box' outside the range of scientific psychology. The shift from behaviorism to the cognitive paradigm came with clear evidence that stimulus-response relationships depend on the abstract *information value* of stimulus input. Extensive research led to the conclusion that inside the 'black box' between observable stimuli and responses are information processing functions of discrimination, attention, memory, innate drives, and intelligent decision making historically related to the mind. Initial cognitive models proposed a 'horizontal' linear sequence of stages in which parallel inputs narrow down to serial behavioral output, such as Broadbent's [57] 'filter theory.' Two-process theories proposed an automatized or 'zombie-like' unconscious mode and an effortful, controlled conscious mode of processing. [58, 59] Further research added a 'vertical' component, *depth of processing*, with shallow unconscious and deeper conscious processing:

$$\begin{array}{ccccc} & \text{sensory} & \text{long-term} & \text{response} & \text{response} \\ \text{S} \to \text{senses} - & \text{memory} - & \text{memory} - & \text{selection} - & \text{output} \to \text{R} \\ & \downarrow & & \uparrow & \end{array}$$

central processing, short-term memory, conscious attention

Figure 2. This chart depicts the change from a horizontal model of functional stages of processing to the addition of a vertical component of depth of processing.

Similar two-process theories were prominent before the behaviorist paradigm rejected unobservable subjective processes. These historical introspective theories delineated focal conscious attention and a peripheral fringe. [60, 61, 62, 63] The fringe region can be viewed as including both shallow input processes and an interior fringe *deeper* than focal attention. This is somewhat implied in contemporary *global workspace theory (GWT)* [64] in which consciousness is a generalized workspace analogous to center stage of a theatre with peripheral cognitive processes surrounding it, akin to a holistic conscious self.

Theories of a conscious fringe deeper than focal attention are relevant to an important debate concerning the primacy of either cognition or affect. One side argued that emotions depend on cognitive evaluations in attributing meaning to events. [65] The other side argued that cognitive evaluations depend on deeper undertones of conscious affect that are more powerful contributors to the intentional direction of behavior than conscious cognitive thinking. [66]

The term *affect* includes both emotions and feelings. To *emote* is to express subjective feelings in objective behavior. To delineate them, 'emotions' are observable behavioral expressions of psychobiological processes in the body; and 'feelings' are *inner felt senses* that are unobservable, deeper interior mental processes. [67]

Cognitive processes are deeper when comparing *cognition* and *emotion*; affective processes are deeper when comparing *cognition* and *feeling*. Affect as feelings deeper than cognitive thinking adds to the model of levels of depth. Conscious processing has a vertically deeper inner fringe of feelings, including a 'felt sense' of self. The human information processing system as a unitary self is reactive to *bottom-up* input, but also proactive and directed by *top-down* feelings and thoughts to achieve *intentionally* its own valued internal and external states. The model below depicts shallower lower-order sensory functions on the top and deeper higher-order functions on the bottom. It summarizes a massive body of research in the past 150 years not mapped into a general model of human mind: [68]

Sensory Environment
Input (S) →   → Output (R)
↓   ↑
Sensory receptors   Behavioral effectors
Sensory memory
Long-term memory
Peripheral conscious attentional fringe
Short-term memory (attention-activated long-term memory)
Focal conscious attention
Conscious fringe of deep feelings
Conscious inner sense of self

Figure 3. This chart depicts the functional model of mind developing in cognitive research in scientific psychology, but not pieced together into a general model.

The functional model depicted above is reasonably consistent with the model of levels of mind on the subtle non-physical level in Sankhya, as shown below (Vedic terms are on the right):

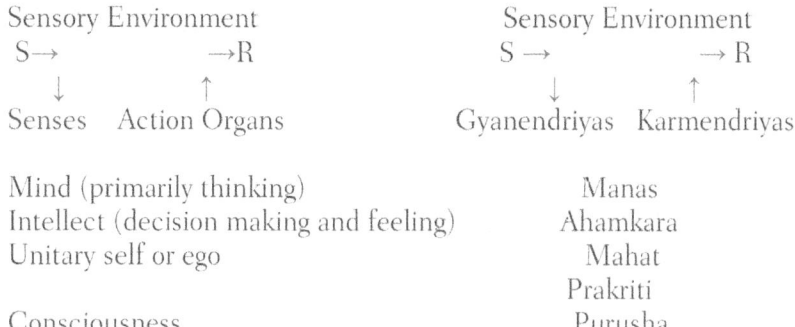

Sensory Environment   Sensory Environment
S→   →R   S →   → R
↓   ↑   ↓   ↑
Senses   Action Organs   Gyanendriyas   Karmendriyas

Mind (primarily thinking)   Manas
Intellect (decision making and feeling)   Ahamkara
Unitary self or ego   Mahat
                                 Prakriti
Consciousness   Purusha

Figure 4. This chart shows the correspondence in the functional model of mind in cognitive research that has developed over the past 150 years with the ancient model as articulated in the Sankhya model of levels of mind. It is a section of the chart in Chapter 5.

Similarities in the models are apparent. However, of particular importance is that in the Sankhya model consciousness underlies all the levels of sensory, cognitive, and affective processes of the individual self *both* functionally and structurally. This means that individual mind is not just in the brain. Rather it is held to be a nonlocal mental space underlying the physical, consistent with some cutting-edge quantum theories as noted earlier—underlain and permeated by universal consciousness or Being (Purusha) akin to unified field theory. This sharply contrasts with the Freudian model still prominent in psychology of the unconscious as underlying conscious mind, which is based on classical physical theory. We will now outline each level of mind in the Sankhya model.

In ancient Vedic science the limiting of the simultaneous wholeness into the phenomenal sequence of levels is sometimes described in terms of five coverings, bodies, sheaths, or filters—called *koshas*. [69] They can be described as functional structures at different levels from subtle to gross that construct individual psychoarchitecture and the phenomenal reflection of individual consciousness. These forms, structures, sheaths, or bodies localize consciousness into a channel of individual experience. Each level reflects increasing limitation and concretization.

In the other direction, from gross to subtle, each sheath reflects increasing expansion, permeability, dynamic liveliness, and integration of energy and intelligence. The most expressed sheath is built of matter particles that exhibit solidity and the least amount of inherent intelligence—where energy and intelligence appear the most independent of each other and the most inert. The increasingly subtle sheathes reflect higher degrees of integration of energy and intelligence toward the infinite integration of energy and intelligence in the unified field.

The most expressed sheath is called *ananamaya kosha,* associated with the physical form or 'body made of food,' built of the five fundamental elements or mahabhutas—a structure made of *matterstuff,* quantized particle interactions that make up the gross level of nature. It is commonly associated with restricting sensory experience almost entirely to the macroscopic gross relative domain.

Subtler than this most expressed level is *pranamaya kosha,* associated with the vital breath, the inward and outward pulsation or reverberation of nature from universality (inward) to individuality (outward), and vice versa. This level can be associated on the gross level with respiration. It also can be understood in terms of nonlocal wave dynamics, pulsations or vibrations of energy and intelligence of the subtle essence elements or tanmatras forming subtle objects of sense—sort of a more abstract structure or form made of *energystuff.*

Gyanendriyas (senses) **and** *karmendriyas* **(action organs).** Information from the environment comes in contact with the outer sense organs—ears, skin, eyes, tongue, and nose—according to 'billiard ball' and particle interaction models of causality. The information is transformed into neural processes in cortical and sub-cortical brain areas, and at some point become subjective hearing, touch, sight, taste, and smell in the inner organs that mind, heart, and ego use in conscious experiences. These inner senses are subtle *gyanendriyas (gyan* refers knowledge). The *karmendriyas* are the five organs of action (arms, legs, etc.) in their non-physical counterparts (*karm* refers to action). In this model of the inner dimension in Sankhya, the five senses refer to the subjective functions of hearing, touch, sight, taste, and smell.

The senses connect the outer environment to the deeper inner level of the thinking mind. Scientific research has progressed quite well in understanding the early stages of input through the physical sensory receptors to corresponding brain mechanisms. But the deeper that the structural analyses go into the objective brain, and especially the functional analyses into subtler subjective levels, the more difficult it is to research sensory, perceptual, and cognitive functions.

The inner levels of phenomenal experience are not observable from the outside, making deeper levels of perception, thinking, and feeling quite difficult to research using outer, third-person objective means. Computers serve as good physical models to help build theories of subjective functions, but they don't address inner *subjective* experiences of sensations, thoughts, and feelings. Later we will discuss how the physical body and matter might interact with subjective senses and deeper mental functions.

Manas and thinking. The word *mind* is frequently used to identify all inner subjective functions. And in this sense, it contrasts with the word *body*, which includes the nervous system and brain. But when the word *mind* is used as distinct from the other inner levels of senses, psychological heart, and ego as in Sankhya, then the mind as *manas* refers to a specific level of inner mental functions. In this meaning, the inner level of mind carries out the tasks of processing sensory input, attending, thinking, analyzing, and reasoning. It also guides the organs of action (e.g., arms and legs in behavior). The mind is the manager of both sensory input from the environment and behavioral output into the environment. The outward and inward flow activates various layers of depth in the mind.

For example, when the senses get information from the outer environment of potential danger, an instinctual reaction such as the fight-or-flight response may get triggered. This results in automatically putting into effect a set plan of action for self-protection. This set plan of survival action has been built in through our genetic history, and modified through past learning.

The deeper level of mind associated with active thinking rather than automatized processing, is the level that analyzes, dissects, categorizes, and conceptualizes, in order to understand with clarity and precision to direct more effective action. Sometimes the phrase *screen of the mind* is used to describe the part of the inner dimension that puts together perceptions, memories, and thoughts into plans about how to act.

At times it may seem like there is kind of an inner screen—sort of like a movie screen or film screen—that displays perceptions in the mind. For example, when we have a clear memory of something, it is almost as if we can view it again in our mind. This is an inner experience on the screen of the mind. It can be associated with the same screen of the mind that dreams seem to be projected onto during sleep.

This level of manas or mind is associated with *manamaya kosha*. It is the level or form through which very subtle energy and intelligence manifest in thoughts and mental impressions. It can be described in terms of pure mind—sort of a subtle structure or very abstract substance made of *mindstuff.*

The decision making function of the level of mind is also frequently associated with the term *intellect*, also called *buddhi* in Vedic terminology. As deeper levels of the mind (manas) are needed in problem solving and decision making, the inner level referred to as the discriminative intellect is activated. Discrimination can be understood as comprising two processes: *dividing* and *uniting*. It involves a basic duality of things fitting together and things fitting apart, in creating categories, concepts, theories, and more all-inclusive models, paradigms, and worldviews. In a general way, the dividing or differentiating part can be associated more with functions on the level of mind (manas), and the unifying part can be associated more with the functions of feelings (and ahamkara), both relating to the intellect (buddhi). However, frequently the analyzing or dividing aspect of intellect is emphasized as the discriminative function.

Ahamkara and feeling. Even deeper than the level of thinking (manas) is the feeling level, which can be associated more with *ahamkara*. It can also be associated with the concept of the psychological heart, not just the physical heart in the gross body. This level provides the psychological energy or *will to act*. To contrast feelings with thinking, the deeper level of feeling is less interested in differences, conditions, distinctions, boundaries, limitations, and restrictions as is the level of mind (manas). It is more involved in uniting, harmonizing, integrating, and connecting together objects and experiences. This aspect of the inner dimension is intimately associated with the concept of the individual self as a unitary individual experiencer.

The concept of individuality which defines individual self is intimately associated with the uniting and dividing functions of intellect, in terms of the discrimination of one's own *self* as a unitary wholeness separate from everything else—from things *other* than one's own individual self. It is also involved in unifying all experiences into the individual self, the individual unitary ego.

This level of the inner subjective dimension is associated with *gyanamaya kosha* and the concept of pure knowledge, the subtlest differentiation between the universal value of intelligence and individuality. It is a very abstract form made of impulses of intelligence, and associated with the level of intellect—sort of *intelligencestuff*.

Mahat and being. In Sankhya the subtlest level of the individual is the level of the unitary self, the experiencer, or the "*I*," which generally can be associated with the term mahat. In western psychology it is frequently associated with the concept of 'ego.' This level integrates the senses, mind, and intellect into a holistic inner individual experience. It is sometimes attributed to be the experiencer of the changing values of actions, sensations, perceptions, memories, thoughts, and feelings. It is the sense of '*I am*' that underlies the sense that '*I feel,* '*I think,*' *I perceive,*' *I act.*' It can be associated with the sense of agency or performer of actions in ordinary waking.

This meaning of ego is not the same as the meaning implied in terms such as 'egotistical' or 'egoistic.' The individual ego can be thought of as the container of individual experiences, not just a collection of them. As used here, individual ego or self is the most fundamental and integrative level of the inner subjective dimension. It is the level that seems to have the phenomenal feeling it is acting out the screenplay or *film* on the screen of the mind. In a sense it can be thought of as the most phenomenally real individual level.

The individual self or ego is also the level at which the dual qualities of unifying and differentiating of the thinking mind, discriminative intellect, and feeling psychological heart have their deepest and subtlest individual expression. The inner organs of action and sense, thinking mind, feeling intellect, and individual ego or self comprise the inner subjective dimension of the individual *soul*, sometimes called the *jiva* in Vedic literature. [5]

Wherever there is the sense of self as opposed to other, whenever the self identifies itself as a separate individual, then there is the possibility of pulling into itself, or pushing away from itself. There is the possibility of inward and outward, of accepting and rejecting, of receiving and sending, of giving and taking, of having and not having, of attachment and non-attachment, of love to unite and fear of separation. All the phenomenal processes of existing, being, wanting, feeling, thinking, sensing, and acting as an individual emerge from the dynamic interaction of these two apparently opposing forces.

When we carefully consider what tells us deep inside that a point of logic is correct, it comes down to an underlying feeling we have deep inside that it makes sense, it *feels* right on a very deep intuitive level inside. Even in a non-emotional decision such as a point of mathematical logic, at its basis is the level of feeling or inner sense about it.

The levels of thinking, feeling, and ego or self are not completely distinct levels. They are ranges of inner experience where the qualities of mind, heart, or self are most prominent. The intellect (buddhi) can be viewed as extending from the level of mind to the subtlest level of ego or self, in the sense that differentiating and unifying extend throughout all these levels. These distinctions reverberate through all levels of nature, and for example can be taken further to the distinctions of Prakriti/Purusha, diversity/unity, point value/infinite value, self/Self, and infinite dynamism/infinite silence.

Intimately related to this level of the inner subjective dimension is the subtlest structure or form called *anandamaya kosha*. It is said to be the almost perfectly balanced state of the three fundamental forces (gunas of sattva, rajas, and tamas) in the holistic three-in-one Vedic model. It is associated with bliss (*ananda*) overlaying (through

Maya) the level of consciousness itself—the almost universal value of individuality. This kosha can be associated with the concept of pure individual being or individual ego, the individual reflection directly contacting universal Being. It is the subtlest, most abstract level of individuality, sometimes described as the bliss-filled body—sort of pure amness or *egostuff* that is of the nature of bliss.

This encouragingly suggests that the most fundamental aspect of individuality is the experience of pure bliss. Underlying this levels of the inner dimension is said to be consciousness itself, pure universal Being associated with the concept of the unified field.

To summarize, in the Sankhya enumeration the unified field of nature (Purusha/Prakriti) first expresses itself through limiting the infinite eternal universal Self or universal consciousness into the finite subjective level of cosmic ego or cosmic Self within which are individual selves. At this level, the point value of the infinity/point duality begins to be expressed as individual self in the duality of self and other.

This can be said to be expressed further via additional limitations into the other levels of the subjective inner dimension (intellect, mind, senses). It is at the level of the intellect (buddhi) that the discrimination of self and other becomes more expressed, into the observer (manas), process of observing (senses) and their objects of sense (in this case the subtle objects (tanmatras). Emerging from the subtle objects of sense are the gross objects of sense (mahabhutas). The same qualities, ingredients, or forces of nature of sattva, rajas, and tamas, when projected to the physical level, are expressed in terms of information, energy, mass and the maintenance, creative, and destructive operators described in previously.

In this chapter we have outlined how in the ancient Vedic model the gross objects of sense are said to condense from the subtle objects of sense (tanmatras). In the next chapter we will discuss how these levels relate to subtle and gross bodily forms in the model.

Chapter 9

Gross and Subtle Body

The Sankhya system can be interpreted as indicating that the physical body is built of the gross constituents (mahabhutas). The subtle individual soul is comprised of the nonlocal, non-physical essences (tanmatras) along with other levels of the inner subjective dimension including indriyas (organs of sense and action), mind (manas), psychological heart/intellect (ahamkara), ego or self (mahat), and also the unified field of consciousness (Prakriti/Purusha) or universal Self.

Physical body and non-physical soul

According to this model of levels, the individual soul can connect to a gross physical human body composed of the mahabhutas. But it also can exist in subtle form without connection to a gross physical body. In this sense the individual soul relates to a *subtle body* associated with the tanmatras and the other subtler levels.

Because the principles of thermodynamics and particle interaction model of causality don't apply at the nonlocal subtle level in the same way as the gross level, the subtle body does not need digestive, respiratory, circulatory, muscular, glandular, neural, or skeletal systems to move energy around like in the gross body. In this subtler level, objects inherently reflect more energy and intelligence, so to speak relatively more self-luminous and self-sentient, rather than phenomenally inert as with physical matter. It is not subject to the limitations typical of objects within the gross relative field. This means experience is not via electromagnetic mechanics and not limited by light-speed and the ordinary spacetime gravitational field.

However, the subtle levels can interact with these grosser levels even though not subject to their limitations. They can be viewed as refined levels permeating gross levels but not restricted to them. Going from gross to subtle does not mean losing specificity, and it does not mean losing the ability to interact with the gross level of nature. Nonlocality does not necessarily mean *not* local at all.

136

Consistent with the Sankhya account, when connected to a gross physical body built of the mahabhutas the individual soul functions in the ordinary local physical world of independent objects as a living biological organism. When the gross physical body of a living organism stops functioning, only the physical body is said to end and dissipates according to entropic thermodynamic processes characteristic of inert matter ('ashes to ashes and dust to dust').

But the individual soul continues to function on the subtle non-physical level, ordinarily not available to experience in the gross physical level. In this sense our ordinary conceptions of life and death relate to whether the individual soul is connected to a functioning gross physical body or not. According to the model, the individual soul is not annihilated when the physical body stops functioning, as in the classical reductive physicalist view and its derivative materialistic philosophies. If the senses are not sufficiently refined in their functioning, however, they are restricted in the physical body to sense only their corresponding objects on the gross physical level.

As sensory abilities are more developed, the subtle objects of sense composed of the tanmatras can be sensed even when connected to a gross physical body. In this view the universe is much more expanded and vaster than the ordinary observable or visible levels. There are many additional subtle layers that cannot be sensed through gross sensory functioning, which naturally become available in subtler, more refined higher stages of development.

Layers of the subtle level of nature can be envisioned as analogous to the vast range of frequencies in the electromagnetic spectrum that can be tuned into and that each has its own world of sounds and sights like television channels. In a way similar to the layers in gross ecological kingdoms of plants, insects, animals, and humans, there are subtle levels.

In this view the individual soul is born or embodied into a physical body (built of mahabhutas), occupies it, and lives in the ordinary gross physical world for its biological life-span. When the physical body wears out or can no longer function, the soul sheds it—like worn-out clothing—and continues in the subtle levels of nature in the subtle body. After some time the subtle body again links with another gross physical body, as a newborn infant. Belief systems that hold there is only one biological lifetime can be viewed as largely accurate portrayals that emphasize the physical level and characterize shorter-term phases of larger evolutionary cycles in nature.

According to the Vedic perspective, it is not that survival behavior, intentional top-down causation, self-awareness, and consciousness are *created* as emergent properties in the process of biological evolution in the physical—which did not exist before, have no purpose, or don't exist at all. In contrast, through long periods of time they become expressed in higher stages of evolutionary development. And in the more advanced human species, they are increasingly prominent in natural evolution toward full potential. In this view, the individual soul does not magically get created at the time of conception, but rather links to the individual gross body when the body is developed enough to house it. This link ends when the biological organism of the individual wears out or is lethally damaged.

It is not that more complex systems and higher species of living organisms inexplicably and magically appeared in biological evolution, as in the reductive physicalism of mainstream modern science which doesn't account for subtler processes underlying the physical. Rather, the evolutionary processes all along are impelled according to deeper and more fundamental causal dynamics within the holistic self-referral model of the consciousness-mind-matter ontology.

In this context the concept of emergence refers to higher expressions of latent functions, not emergent phenomena with no ontologically real substrate or that are just epiphenomenal. Descriptions of these higher evolutionary processes appear throughout the history of religious and spiritual traditions—although the descriptions are quite obscure in many of them. With advances in modern science, hopefully these subtle levels historically so difficult to describe in religious traditions can be seen to fit into a coherent scientific model.

Phenomenal nature is said to be an eternal cyclic process of manifesting parts of nature, and evolving back to the ultimate wholeness inherent in and progressively enlivened in the parts. It is held to be a never-ending self-referral process across vast eons of time that can be said to extend from phenomenally non-conscious, inert parts of nature to the ultimate wholeness or totality of the infinite eternal unified field of universal Being. The theory of evolutionary biological emergence of more advanced species can be viewed as fitting in as one phase of this completely holistic process.

In other words, higher-order, top-down mental processes emerge in the physical associated with increasingly complex physical structures, consistent with reductive physicalism. But these complex physical structures don't create mind and consciousness. Rather, guided by subtle mental processes all along, they allow these latent higher-order processes to become expressed. While this holistic account is generally consistent with the reductive physicalist account, it is a much bigger picture of the full range of human experience and the full range of natural evolutionary processes.

Mind/body connection. To summarize the holistic picture of levels of nature overviewed in this book, all gross levels can be understood to be emergent from an underlying matterstuff associated with the relativistic spacetime gravitational field. This also can be associated with the *ananamaya kosha*, the physical form of the individual body.

This subtlest of the gross level of nature corresponds in Sankhya to the mahabhuta of space (akasha). This level is an emergent property of the grossest of the subtle level of nature, associated with the tanmatra of the essence of earth. This tanmatra contains all the five constituents expressed in sequential symmetry breaking into the five gross constituents. The five subtle tanmatras can be associated with *pranamaya kosha*, the form of the individual associated with the flow of subtle energy not yet defined in modern physics.

Each gross constituent thus contains in it the subtle tanmatras, as well as each even subtler level of the inner dimension of subjectivity, again including senses, mind, intellect, individual ego or self associated with the other three koshas, and individual and universal aspects of consciousness itself. All of these levels ultimately are nothing other than the infinite eternal unified field interacting within itself as the basis of all subtle and gross phenomenal levels—the completely holistic three-in-one Vedic model of the totality of nature.

Aspects of Vedic literature further include various descriptions of the connection between subtle non-physical and gross physical bodies. Here we will discuss one interpretation of this relationship. And then we will consider another level of detail, but not a comprehensive explanation as in the totality of Veda and Vedic literature.

Phenomenally, at some point there is a transition from subtle dynamics to gross mechanics. In the context of physical theories, there is a basis for the interpretation proposed earlier that this transition takes place at about the level of the Planck scale. A reasonable interpretation is that at this stage of manifestation, the abstract and more extended finite field of the tanmatras can be said to condense or precipitate into Planck-scale quantization with the increase in tamas (the entropic, concretizing or resisting change influence) limiting the corresponding rajas (creative, activating, or propelling influence) and sattva (the maintaining, negentropic, attractive or unifying influence). The interactions of these abstract qualities or influences phenomenally make more expressed the point value of independent objects.

And their relative strength might spontaneously structure Planck scale quantization from which emerge the fundamental force particles at larger time and distance scales of the gross level of nature. Change on the gross levels of nature with particle interaction causal mechanics of independent objects (mahabhutas) naturally takes place through inherent dynamism (energy), inherent resistance to change (mass), and inherent order (information) in the known laws of nature. But further, these processes also are influenced by the nonlocal causal wave dynamics (tanmatras) of the subtler inner dimension from intentions of nonlocal minds.

Loci connecting subtle and gross bodies. The Vedic literature includes different perspectives on the relationship between phenomenal mind and matter. One of the perspectives will be used to describe another level of detail about the relationship between the subtle and gross bodies, [70] which seems appropriate to include as we speculate about future research on the age-old mind-body problem based on scientific advances overviewed in this book.

140

Within the context just described of all levels of the gross body emerging from and influenced by the subtle levels, particular areas or points of connection are emphasized. In this view there can be said to be seven key loci in the subtle body that have their correspondence in the gross body. Sometimes called chakras or marmas, they are described as key centers of energy that connect and transition between subtle and gross.

These loci are said, according to the model, to correspond generally to points along the physical spinal column of the gross human body, from the base of the spine to the reproductive area, abdomen, heart, throat, forehead, and crown of the skull. Each of these loci are said to have significance in the process of full development to higher states of consciousness and a completely unified understanding and experience of nature. Of course extensive research is needed to evaluate the efficacy of this very subtle model of the connection between nonlocal mind and local matter/body. Subtler epistemological approaches from Vedic literature have the potential to enhance systematic means of gaining knowledge for much deeper integration of theoretical knowledge and empirical validation.

The Sankhya model provides a structure that is ontologically rich enough to accommodate causally efficacious mind. It also provides a basis for understanding how subtle experiences historically associated with spirituality and religion that seemed impossible or delusional within reductive physicalism fit into a rational scientific model even more consistent than the mainstream reductive physicalist view.

The integration of science and religion, as well as matter and information or material bodies and 'divine energy,' is progressing in a scientific context, discussed in this book in terms of an expanded ontology of levels of nature and subtler views of spacetime. In the final chapter systematic means to validate the expanded ontology will be discussed briefly in the practical epistemology of Yoga. This model of the means of gaining knowledge emphasizes experiences beyond the physical. This seems to be the stage where many contemporary scientists and religious scholars appeal to hope, [71] without openness to systematic natural means for direct empirical validation in higher states beyond ordinary waking state experiences.

PART III

EMPIRICAL VALIDATION

Chapter 10

Epistemology in Modern Science and Yoga

As noted in prior chapters, experimental validation of levels of nature at the limits of the physical is increasingly difficult, even using our most advanced experimental methods. The immensely significant paradigm shift now underway incorporating holism suggests increasing openness to the possibility of additional systematic methods of validation, briefly introduced in this final chapter.

The main thesis here is that the Vedic tradition applies direct empirical means to gain and validate knowledge that complements indirect objective experimental methods. It also brings a more integrated and personally meaningful value to scientific knowledge, historically missing in the indirect experimental approach. It applies the scientific tenet that nature is orderly to our own minds as part of that order. In this view our minds and the universe we observe with them share the same source and the same laws of nature. This allows knowledge to be educed directly in the 'inner laboratory' of the conscious mind, in addition to the outer experimental laboratory.

We will now briefly summarize how we have developed our understanding of nature in the absence of theorized higher states of consciousness and direct experiences of a richer ontology of levels of nature beyond the physical. Hopefully it will be useful for developing a rational understanding of systematic ancient Vedic means of gaining knowledge to supplement the means we have relied upon in modern scientific epistemology associated with the ordinary waking state.

There has been virtually no recognition of the state dependent limitations of us as scientific investigators within which modern science has been practiced. This is a great loss that has now led to a pivotal time in history in which the outer focus on matter has gotten ahead of inner development of our minds. The consequences are increasingly dangerous. But also there is now growing recognition of systematic means to transform human society into a much more coherent and fulfilling quality of individual and collective life.

Modern scientific epistemology

Epistemology concerns the means of gaining knowledge. Throughout human civilization two major means have been practiced: *reason* and *experience*. Reason relates to the assumption that we have inherent ability to gain knowledge directly through rational, deductive thinking. Experience emphasizes inductive thinking based on direct sensory observation.

The scientific method integrates the logic of deductive reason and inductive experience. Sensory observations provide tangible evidence to develop logical theories predicting how natural phenomena behave. The predictions are tested through careful re-observation; and then the theories are reevaluated for a better fit with empirical results.

In this *self-correcting* process, theories are continually improved to withstand rigorous logical inquiry and empirical testing. When deductive reasoning is not accompanied by empirical validation, the self-correcting process is incomplete. This can be viewed as the issue with respect to concerns arising about contemporary mathematical models that have become quite speculative, with no empirical support and seemingly no available means to validate the speculations.

In modern science, precision is increased by use of accurate measuring devices and protocols; careful recording, categorization, and classification of phenomena; standardized terminology and symbols; formal descriptive and inferential statistics; mathematical modeling in theory construction; and other safeguards to avoid errors. Applying this systematic experimental approach has primarily involved reductive probing of smaller and smaller time and distance scales and higher energy and temperature states.

As mentioned earlier, analyses of theorized objects at unobservable scales increasingly rely on conceptual models of what the object is that is being measured, and what the process of measuring actually means. As research has probed deeper than tangible matter to quantum fields, the *interdependence* of object and observer increasingly has encroached upon scientific objectivity. Matter and mind, object and subject, physics and psychology, can no longer be considered independent, given the major advances to a more integrated scientific account of nature.

Methodological rigor and precision are important for increased accuracy, and to minimize biases. But they don't eliminate subjectivity from science and processes to gain knowledge generally. From a psychological standpoint, sensory experience isn't merely innocent sensory reception of *objective reality* as given to us by nature. Sensory experience is shaped by processes deeper than the senses such as prior beliefs, reasoning, and learning based on accumulated experiences—as well as genetic inheritance that structures how the senses and brain function. A key area of psychological research is the influence of conceptually-driven *top-down* processes of reason on the data-driven *bottom-up* sensory processes.

To further minimize inconsistency, unreliability, and bias in reasoning and sensory experience, the objective approach relies on public agreement among investigators. This critical evaluation process builds inter-subjective agreement about the theories and their empirical support, sometimes called *consensual validation*. Scientific investigations are open for evaluation and replication, associated with the outer third-person perspective. Research is presented in public forums allowing others to evaluate the research and replicate it.

Modern science can be thought of as a democratic process in the sense that, at least in principle, anyone can contribute to and validate the theories and experimental results. Also, theories and evidence supported by the majority of investigators are more accepted.

On the other hand, there are recognized experts whose views receive stronger acceptance than others. In this sense science is also an appeal to authority. It has some similarity—some might say increasing similarity—to religion with respect to the role of authority in establishing generally accepted or orthodox knowledge. And further, it is widely appreciated that *scientific truth* or *scientific facts* cannot be based on democratic consensus but on the authority of nature itself, to the degree that we can observe and comprehend it.

Directly related to this point, it also is important to recognize that scientific knowledge is not based only on sensory experience and logical reasoning. Deep intuitive insights play a fundamental role too. The insights relate to assumptions and premises which undergird scientific theories. These assumptions have not been tested via objective methodology. As working assumptions they have achieved wide acceptance largely on the basis of reason and intuitive-like beliefs—and to some degree on faith in the natural order of things and the inherent ability of humans to reason about and experience it.

These assumptions relate to the concept of a paradigm. Paradigms involve deeper conceptually driven top-down intuitive beliefs that can exert a strong influence on reason and sensory experience. Scientific facts are *theory-laden*, in that the overall theoretical framework gives the facts meaning and significance. In part, what is and what is not observed is a product of the theory that specifies the measurement procedures resulting in meaningful data in the context of the theory.

Likewise scientific theories are *paradigm-laden*. The paradigmatic frame of reference and its pretheoretical assumptions significantly shape what theories are conceived, proposed, and tested. [5] Major paradigm shifts can occur based on new facts, new theories, new technologies to test the theories, and even new insights or transformative ways of viewing the theories and facts. [72] Perhaps the most cited examples are the shift from classical to quantum physics, and the shift from positivist behavioral to cognitive psychology.

Paradigms and related worldviews are products of psychological, biological, sociological, as well as economic and even sociopolitical influences involved with a particular knowledge system. They are relevant not only in modern science but also other knowledge systems including religions. In part paradigms and worldviews—secular or not—are identifiable by their assumptions, intuitive insights and beliefs. And once a paradigm is widely accepted, it can be quite resistant to change. A paradigm shift can evoke strong emotions of fear, resistance, and defensiveness in that it may imply vast amounts of time, energy, money, and reputation in the old paradigm was wasted.

Whether scientific, religious, or otherwise, the basic means to gain knowledge are experience and reason. But also pretheoretical assumptions associated with intuitive-like beliefs play an even more fundamental role. Gaining knowledge involves assumptions about nature and approaches to test them shared by individuals with similar beliefs, reasoning, and ranges of experience. Though we might like to think the process of knowing proceeds from observation to theory to validated theories confirmed by empirical testing, it also proceeds perhaps even more often in the opposite direction from implicit intuitive beliefs to logical theories to empirical validation.

However, the incredible precision achieved between mathematical theories (reasoning) and observed phenomena (experience) in science cannot be diminished, and deserves full recognition. It is a powerful testament to the orderliness in nature that appears to underlie both the objective natural world and our subjective minds we use to examine it. It supports rational faith in the fundamental orderliness of nature.

But objectivity cannot be completely divorced from and independent of subjectivity. The public 'third-person' objective experimental approach in modern science necessarily is also a private 'first-person' subjective approach. Objectivity depends fundamentally on minimizing distortion and increasing orderliness in our subjective minds for more reliability and accuracy in reason and sensory experience, to be able to identify orderly patterns in nature.

It further is important to recognize that many factors influence subjective mental processes through which objective knowledge is gained and worldviews are built. Errors in perception, affect, reason, memory, and intuition can happen. In the same way that a measuring device can malfunction, physiological and psychological processes in the knower also can malfunction due to fatigue, stress, disease, rigid thinking, or other disorderly influences that limit intuitive beliefs, reasoning, and sensory experience. Even our daily lifestyles significantly influence how our bodies and minds function, adding to stress and incoherent functioning or to increasing balance and more integrated perceptual, cognitive, and intuitive functioning.

These factors have a major influence on reliability, consistency, and accuracy within each individual, and also across individuals. It is fundamentally important to recognize that consensual validation is largely defined by the state of functioning of those who contribute to the consensus. Not only must the object and the process of observing be considered in gaining knowledge, but the state of the observers also needs to be recognized as fundamental to what is attributed to be valid knowledge. The state dependent nature of knowledge has been almost completely overlooked, especially in the 'hard sciences.' Unfortunately even many of the most respected scientific authorities and visionaries have not acknowledged and explicitly taken to heart this fundamental issue about their own state of conscious experience. The subjective state of the scientist can no longer be ignored in objective science.

An example of a source of subjective inconsistency across observers is differences in enculturation, especially related to language. As the various forms of cultural and language barriers are bridged, the underlying consistency across individuals is recognized and higher levels can be achieved. Another important source of subjective inconsistency is developmental differences, such as perceptual-cognitive development. At different developmental stages, people experience and reason quite differently. What was initially experienced and reasoned as incorrect, impossible, or even mystical becomes obvious at more developed stages. But even among those with highly developed reasoning and compatible communication systems, different paradigms and worldviews obviously still emerge.

Perhaps the most significant contributions to subjective inconsistency are developmental differences in deeper levels of mind associated with intuitive insights. Consensus is based on the level of functioning of those who contribute to it.

It isn't surprising that the strategy in modern science of emphasizing objectivity and minimizing deeper subjectivity would be reflected in the scientist's personal lives. Reasoning and sensory experience are common functions of the ordinary waking state of consciousness. The entire enterprise of modern science—as well as contemporary views in much of religion and spirituality—is based primarily on intuitions, reasoning, and sensory experience shared by individuals in ordinary waking experience. The level of functioning of the mind—most fundamentally the *state of consciousness*—shapes the consensus and the intuitive beliefs, reasoning, and sensory processes.

The ordinary waking state of consciousness is characterized by an experiential gap between outer objective and inner subjective. This gap is the basis for the assumption of object-subject duality in classical science. It is a fundamentally fragmented experience, contributing to fragmented worldviews. This gap needs to be bridged for unification of all the laws of nature that science rigorously and relentlessly pursues, and that religion also has long pursued with its own language and methods. [23]

Consistent with the incredible match that has been achieved between scientific knowledge and empirical phenomena, if higher states of consciousness can be systematically developed it would promote even higher consistency and accuracy of knowledge. More integrated understanding and experience of nature and our place and role in it could be achieved. Fortunately, and with great significance, this is now finally emerging on the forefront of modern science.

Modern science is now at the doorstep of the ultimate unification in unified field theory, truly a profound achievement. [3, 5] It is significant that the general concept of a single underlying source of everything in nature—apart from its specific descriptors and cultural icons—has been perhaps the most widely accepted intuitive belief across religions. Unified field theory has brought modern science to the doorstep of a unified view of nature that religions long ago intuited in their own language and context.

The holistic unified field-based account can revolutionize popular understanding about the place and role of human life in nature. When unified field theory and its implications are actually taught in our educational systems, it can reverse disintegrative trends from cultural relativism, fundamental randomness and nothing, and existential meaninglessness that have been eroding the coherence and integrity of society.

But modern scientific progress toward ultimate unity is in terms of theoretical understanding—intellectual wholeness. It has not yet included *direct empirical validation* of unity. Unification remains a most daunting task. This is evident even in the most successful theories in modern science, which though accurate as far as they can be tested in the current paradigm, still reflect a deeply fragmented account of nature. Applying the objective approach to everyday life, our inner subjective experience has remained separate from the outer objective world. The primary locus of experience remains the concrete, sensory, external, objective material world. The objectified means to gain knowledge allowed us to progress a long way out of superstition and ignorance. Along with its many successes in technological applications, unfortunately it also is associated in recent years with lack of fundamental grounding that has rendered daily life devoid of meaning for much of society. Focusing only on the gross outer level of nature, modern and post-modern life has been tightly bound to the superficial flatland of material existence.

Even the most successful theories in modern science have contributed to popular socio-cultural views and lifestyles associated with flatland materialism. For example relativity theory was widely interpreted as supporting views that there is no absolute objective or subjective grounds for moral behavior; everything is relative. Also quantum theory was interpreted in terms of the devaluation of life based on the fundamental randomness and thus the meaninglessness of life. It has contributed to the notion that the world is not real, just a phenomenal pattern of abstract wave fluctuations in fundamentally random intangible fields that have no inherent meaning.

This notion even has been associated with interpretations of ancient views that the universe is unreal and illusory. But what is left out in considering ancient views is the developmental context beyond sensory experience in the ordinary waking state to higher states in which the full range of levels of nature is considered. This context is crucial in order to understand the meaning of the relative finite world as illusory in ancient traditions.

Evolutionary theory similarly was interpreted in terms of random variations in genetic material with no direction or purpose other than meaningless biological survival. But if in evolutionary theory biological evolution is due to random genetic drift underlain by inert random quantum fluctuations in a closed causal nexus, there is no room for survival to become *favored* in nature in the first place. Evolutionary biology assumes biological survival somehow becomes preferred and is a fundamental motivating factor in natural selection. But there is no place for this value to be inserted into the closed causal physical nexus that is believed to be completely *valueless*, according to the current orthodox model in mainstream modern science.

Interpretations of these most successful theories in modern science have contributed greatly to a deep tear in the psychosocial fabric of modern and post-modern life. With these views, for the most part science and spirituality have been increasingly antagonistic to each other; and religious beliefs have been disdained as naïve and outdated. This has been accompanied by increased psychological angst of the futility of life—fueling reactionary hedonism and overt hostility that is growing today. Doubt was further amplified about whether anything has intrinsic value, which has fostered uncertainty about the basis for ethics and morality throughout human society.

Leaders in the sciences, humanities and arts have made great contributions to the enrichment of society. But the fragmenting reductive and deconstructive views reverberating throughout society from the halls of science in the past century have led toward an existential dead-end, with progressive decline in societal coherence and values. Reflecting on this situation, Maharishi points out that the incomplete objectified approach in modern science and the fragmented educational system based on it: [41]

> "...Inspires the seeker of knowledge to focus on isolated areas for many years... Reveals that in the pursuit of knowledge whatever knowledge one gains, that knowledge itself reveals that there is yet more knowledge to be known.... This makes the path of knowledge endlessly long, and fulfillment out of reach.... [It] Searches for knowledge outside of the Self; thus the knower runs after knowledge and exhausts himself without reaching the goal of complete knowledge.... Failing to bring fulfillment, it is destined to be surpassed by a more complete science.... [It] Creates a society characterized by problems, crime, stress, sickness, unhappiness, and disharmony.... [It] Bestows upon the superpowers the ability to destroy life on earth...." (pp. 192-195)

Indeed, the reductive physicalist worldview can be seen as having massive negative consequences throughout the 20th Century. It has spawned popularist sociopolitical philosophies that can be viewed as having wrought more destruction in just this past century than the entire terrible legacy of wars based on religious fanaticism.

It further is spawning aggressive efforts to create man-machine cyborgs and genetically engineered posthuman life forms, based on fragmented reductive physicalism. Though some developing technologies are obviously quite useful for specialized prosthetic applications, attempts to enhance human abilities artificially overlook the inherent potential for natural human development.

With general lack of deep inner development in both secular and non-secular approaches, we have gotten locked into fragmented gross physicalism almost totally divorced from holism and the natural order of life, including human life. We have 'physicalized' our minds as just 'material bodies,' and manipulate the body as if it were nothing more than a piece of machinery. This has the dark potential to destroy its subtle functioning, such that the natural ability to transcend the limitations of ordinary thinking could be lost—a profound existential danger to humanity based on fragmented, coarse-grained understanding and experience of nature.[73] As Maharishi [74] has stated:

"Those whose hearts and minds are not cultured, whose vision concentrates on the gross, only see the surface value of life. They only find qualities of matter and energy.... They do not enjoy almighty Being in Its innocent, never changing status of fullness and abundance of everything that lies beyond the obvious phase of forms and phenomena of matter and energy, and of mind and individual... Pure Being is of transcendental nature because of Its status as the essential constituent of the universe. It is finer than the finest in creation; because of Its nature, It is not exposed to the senses which primarily are formed to give only the experience of the perception of the mind, because the mind is connected for the most part with the senses." (pp. 24-25)

Fortunately, important progress beyond physicalism is now being made in modern science. But it has taken a relatively long time to gain sufficient understanding to go beyond it, given ordinary sensory experience and reasoning. Some investigators have become enthralled with intellectual rigor, puzzle solving, wonder, cynicism, and sometimes existential empathy and compassion, resigned to the 'reality' of life as ephemeral, random, and ultimately meaningless.

At some point the reductive, objectifying intellectual mind overshadowed by surface materialism paints itself into a corner and finds nothing. This is evident for example in the consensus cosmological theory, as well as views in cognitive science that mind and consciousness can be eliminated entirely from scientific models. Eventually this approach is recognized to be fragmented, incomplete, inconsistent, outside oneself, and unfulfilling. And eventually real mind is recognized as necessary to examine whether anything at all exists. This brings up the Vedic developmental approach of Yoga as an expanded epistemology to gain reliable knowledge and experience.

Yoga epistemology

The holistic Vedic approach begins with the unified field as the field of total knowledge. That is what the term Veda means. [41] It is said to unfold sequentially the parts of knowledge within the ultimate wholeness, eventually identifying wholeness in every part.

Full appreciation of the completely holistic view of nature is said to require development of higher states of consciousness, which include refinement of perception to experience subtler levels of nature. Such experiences are said to be natural, but currently beyond the mainstream scientific consensus restricted to understanding and experience typical of the ordinary waking state of consciousness.

It is quite challenging even to envision the concept of nonlocal mind permeating the physical brain, given an ingrained reductionism, limitations of the ordinary waking state, and lack of direct experience of subtle nonlocal levels of nature. Nonlocal mind does not mean that the ordinary experience of locality and physicality of objects is lost, but rather deeper, more expanded experiences naturally arise.

Ordinarily it feels like our mind is behind our eyes, perhaps in the brain somewhere, as a localized arena of subjective experience or 'screen of the mind.' In higher states, this 'screen' is refined to include much subtler, inherently more enlivened essences of sensory objects, and our inner sense of self is the background observer of the 'screen.' The screen is experienced as presenting subtle as well as gross objects to the individual self. But eventually the individual self is experienced as an unbounded, nonlocal observer of all individual experience.

And further, in the highest state of unity consciousness all objects of experience on any level of nature are directly experienced as essentially the universal Self or pure Being. This means that our daily experience is then not restricted to a view of nature from a localized inner sense of individual existence divorced from the universal totality of existence. But these purported higher states are so far beyond the ordinary waking state that attributes 'self' to a localized area centered somewhere in the brain that even the notion of nonlocal mind can be hard to conceptualize—without at least some transcendent experiences beyond sleep, dreaming, and ordinary waking perception.

Phenomenal experience of subject-object independence is commonly associated with the sense that the objective outer world is somehow more *real*—more substantial, reliable, and consistent—than inner subjectivity. In this objectified experience, inner subjectivity is more variable than the outer natural world. The objective world seems to remain much as it is whether we are in one mood or another, asleep or awake, experiencing it or not. In the constantly changing experiences of the three ordinary states of consciousness, subjectivity varies significantly. In this sense it appears to be less stable and reliable than the outer objective world—absent a deeper foundation of understanding and experience.

In ancient holistic accounts this view of nature characterizes the state of ignorance. The belief that there is an objective world separate from inner subjectivity, and thus that nature is fundamentally dualistic, is identified as a stage of understanding associated with empirical experience typical of ordinary waking consciousness. This experience is held to be a product of the discriminative property of the intellect without sufficient grounding in unity. In Vedic terms it is identified as 'Pragya aparadha,' the *'mistake of the intellect.'*

In this worldview the belief has been increasingly popular that the intellect is the most reliable tool to gain knowledge. Logical reasoning has been the primary means of gaining knowledge in this Age of Science, especially in recent decades when direct sensory observation has been far surpassed by indirect research that relies even more on conceptual models based on mathematical logic. It also has been the primary focus of training in modern education; and certainly is a precious faculty. But while the discriminative intellect has cut deeper into nature with incisive analysis, it also has chopped it into meaningless random bits of matter/energy/information.

Fortunately there is growing recognition of wholeness as the basis of the parts of nature. In unified field theory the source of everything is an abstract field of order, opposite of *fundamental* randomness. As this deeper understanding grows, it can reverse disintegrating trends resulting from cultural relativism and existential meaninglessness.

But the profound concept of unity is still within the framework of intellectual understanding—*intellectual wholeness*. Though a huge step of progress in modern science, it is important to recognize that unity cannot be lived on the basis of intellectual reasoning, a mood, an attitude to try to remain mindful of, or an applied social ideology.

At this pivotal point in modern science, it is profoundly important to recognize that major advances toward unified understanding and experience of nature won't come through manipulation of somewhat deeper levels of the outer material surface of nature still based in physicalism—such as bioengineering to alter our natural genetic inheritance, nano-implants to build human-machine cyborgs, or even colonization of *outer* space, in a posthuman era. Although in many cases these research initiatives reflect sincere efforts to address major concerns of humankind, they are predicated on a fundamentally incomplete understanding of the basis of nature.

Substantive practical advances won't come from dismantling nature as if it were merely inert random pieces of matter. When we objectify nature and attend only to matter, we treat everything as if it were just bits of matter—ourselves included. Daily life becomes increasingly fragmented, complicated and stressful. This concern cannot be over-emphasized in our contemporary socioecology.

Rather than continuing in this increasingly disintegrating direction based on untenable physicalist assumptions, the advances will come from subtler, more profound *alignment with nature*. This is said to be accomplished through scientific application of psychophysical laws for systematic natural development in the inner laboratory of the mind of each scientist.

This relates directly to systematic means of gaining knowledge that have been fundamental to the ancient Vedic epistemology of Yoga. Yoga means 'to yoke' or unify. It refers to practical technologies to refine mind and body and to transcend the intellect to unify in direct experience the individual self with the universal Self or Being, the unified field of everything.

Modern science has progressed far beyond directly observable objective phenomena and the indirectly measurable physical level. This has placed more emphasis on logical reasoning deeper than ordinary sensory experience in formulating and evaluating scientific theories. But like sensory experience, reasoning still involves active mentation. Thinking—whether about matter, energy, nothing, the unified field, God, or the Godhead, as well as introspection, self-reflection, or being mindful of some object of experience—tends to keep the thinker in the mental activity of ordinary waking, in which the basic experience is of subject-object duality.

Most fortunately it is now becoming recognized that there are systematic, natural, simple means to validate empirically the ultimate unity of nature. This is said to be accomplished by transcending the apparent dualities that present themselves to the ordinary waking state of consciousness and the intellect that does not transcend the habitual fragmenting and divisive aspect of its nature.

In the 20th Century materialistic and existential views that life is meaningless became widespread—as if entirely disconnected from the universal value of life. Maharishi points out that this occurs when only the indirect, object-based, third-person approach is applied and the scientist remains within the subject-object duality of ordinary waking. The ultimate underlying unity is said to be hidden, due to lack of systematic development of the inner depths of mind. Maharishi [75] explains:

"Being objective in its approach, modern science brings only intellectual understanding about the functioning of the laws of nature. It does not penetrate into the life of the scientist. It does not integrate his personality. He can do some little jugglery here and there in the field of creation, converting this into that and that into this, but he himself is open to all kinds of destructive values because the modern approach to the investigation of natural law does not and cannot enable the scientist to imbibe knowledge and live it in daily life." (pp. 122-123)

Sages from time immemorial have taught that direct experience of unity is not on the level of intellectual reasoning alone. As teacher who has revived the natural ability to transcend duality to the direct experience of unity from the ancient Vedic tradition, Maharishi has pointed out simply and profoundly: [42]

"Transcending thought is infinitely more valuable than thinking."
(p. 444)

Empirical validation of unity through refining body, mind, and intellect is the purpose of Yoga. Although a few scientists have spent time on both development of intellectual understanding and also transcendence of the intellect, many have not and thus doubt that direct experiences beyond the ordinary discriminative intellect to the non-dual state of unity are even possible. In the following quotes, Maharishi addresses how the direct experience of unity or Being can be repeatedly missed: [76]

"When a flower is seen, then only the flower remains in the mind, as if the mind had been completely annihilated, void of its own glory, and the glory of the flower had overtaken it—as if the flower had overshadowed the glory of the mind itself. The experiencer is missing, only the sight remains and the object... This is called objective life, material life. Matter remains dominant...." (p. 294)

"Being is not appreciated by the mind, although It is its very basis and essential constituent.... The great dignity, the great splendor and grandeur of Its...omnipresent nature is present in man as the basis of ego, intellect, mind, senses, body, and surrounding. But it is not obvious; It underlies all creation... It is the omnipresence of Being that is responsible for hiding Being behind the scenes... It lies out of the realm of time, space, and causation, and out of the boundaries of the ever-changing phenomenal field of creation." (p. 25)

"Since Being is of transcendental nature, It does not belong to the range of any of the senses of perception. Only when sensory perception has come to an end can the transcendental field of Being be reached. As long as we are experiencing through the senses, we are in the relative field. Therefore, Being certainly cannot be experienced by means of any of the senses. This shows that through whatever sense of experience we proceed, we must come to the ultimate limit of experience through that sense. Transcending that, we will reach a state of consciousness where the experiencer no longer experiences.... When we have transcended the field of the experience of the subtlest object, the experiencer is left by himself without an experience, without an object of experience, and without the process of experiencing. When the subject is left without an object of experience, having transcended the subtlest state of the object, the experiencer steps out of the process of experiencing and arrives at the state of Being... The state of Being is neither a state of objective nor subjective existence, because both of these states belong to the relative field of life. When the subtlest state of objective experience has been transcended, the subtlest state of subjective experience also has been transcended. This state of consciousness is then said to be pure consciousness, the state of absolute Being.... The transcendental state of Being lies beyond all seeing, hearing, touching, smelling, and tasting—beyond all thinking and beyond all feeling." (pp. 45-46)

Systematic first-person means to gain knowledge in the Yoga aspect of the Vedic tradition is said to facilitate higher development beyond the ordinary waking state. This includes increased subtlety of experience of deeper levels of mind and objects of experience that have been quite difficult to research and model from the outer third-person objective perspective of modern experimental science.

Maharishi has revived from the ancient Vedic science of Yoga systematic means to transcend ordinary waking consciousness to a fourth state of transcendental consciousness. He has systematized this ancient subjective means to gain knowledge in the Transcendental Meditation® technique and has established it in the context of modern scientific technology. Over the past 40 years, it also has become the most extensively researched and validated mental technique for stress reduction and psychological health and development, with 600 or more studies on the results of TM practice now published, about 400 in refereed journals. [77-87]

Subjective experience in ordinary waking is typically active mental attention directed outward toward objects of sense. This can be viewed as in the opposite direction compared to turning inward and transcending mental activity to the unbounded inner silence of pure consciousness itself. It is described as an effortless natural process of allowing the mind to settle down to its least excited ground state—like a wave naturally settling back into the unbounded ocean.

As a simple comparison, included in the ability to run is the ability to slow down, and included in the ability to slow down and run less strenuously is the ability to stand still. Included in the ability to talk is the ability to talk softer, and to be silent. Likewise, included in the ability to think is the natural ability to settle down and relax to the state of inner silence and stillness—to transcend thinking effortlessly to the least excited ground state of the mind.

This process is said to be a natural and inherent dynamical capability of the human nervous system and mind. However, it is subtle and had been relatively rarely experienced, due to habits of ordinary thinking that interfere with this natural process and render it seemingly quite difficult. The benefits of systematic, reliable transcending apparently have not been widely experienced, either in secular or non-secular traditions. The focus largely has been limited to the range of experiences characteristic of the ordinary waking state of consciousness in the vast majority of the world population.

Maharishi has pointed out that there has been a general lack of understanding of how the mind effortlessly settles down. In *trying* to still the mind, the typical experience is that it is fickle and shifts from one object of experience to another. Long traditions have been established based on the view that the mind then must be controlled to attain inner stillness, and that this process must be difficult.

Methods based on this view attempt to still the mind by applying forms of either *contemplation*—reflective thinking about some idea or object of attention—or *concentration*—effortful focus on a particular object such as an image or the breath. Contemplation has come to mean reflecting on concepts such as peace, inner silence, or the grace of God. Because the mind tends to wander, contemplation is frequently modified to include some form of concentration in attempts to reduce or eliminate intrusive mental content that can distract the mind from being resolute and still—such as focusing on the breath, a visual or auditory stimulus, or a particular idea or concept to hold in attention.

Although perhaps difficult to accept due to long traditions of trying to control the mind, Maharishi emphasizes that the process of transcending is accomplished effortlessly. It involves softer thinking to deeper stages of relaxation in an *effortless* process based on the natural tendency of the mind. The TM technique purposely avoids mental effort and engaging in sensory, emotional, or intellectual processing that tend to keep the mind on the ordinary active surface levels. This natural process is so simple and subtle that it appears to have been overlooked for millennia.

The contrast between the inner silent transcendent state and active mental states is becoming clearer through direct experimental comparisons. Some mental practices correlate with increased gamma synchrony, proposed as the best measurable neural correlate of consciousness. [59, 39] this view is consistent with the ordinary waking state understanding and experience of consciousness as *being aware of* some object of experience. But this EEG pattern is not correlated with reported experiences of transcending mental activity to pure consciousness, which typically involves peak alpha power indicative of restful alertness. [88, 82, 79, 89]

As described in Chapter 1, modern science can be viewed as a *self-correcting* process of gaining knowledge through repeated theory development, testing of the theories, and refining them according to empirical results. But the self-correcting referential loop from theory to empirical validation and back to reevaluation of the theory is not experienced as extending deeper into the underlying basis of reason and experience in the pure consciousness of the knower. It remains on the ordinary levels of mental activity—with the deeper, underlying inner levels of mind remaining an experiential black box, with no direct experience of consciousness itself. Absent of transcendence, the means of gaining knowledge has been restricted to intellectual discrimination and ordinary sensory experiences of the outer objective world, which characterizes modern scientific epistemology.

The natural transcending process in the systematic mental practice of Yoga is said to settle the mind to the underlying source of thought in consciousness itself beyond all contextual limitations. This involves much deeper self-correcting processes beyond ordinary scientific abstract thinking, the result of natural healing mechanisms activated through deep rest in mind and body. This refinement eventually results in spontaneously maintaining transcendental consciousness along with ordinary waking, dreaming, and sleeping, the foundation for permanent higher states of consciousness historically associated with enlightenment.

In this integrated understanding of objectivity and subjectivity, inner development of the knower and the process of knowing are fundamental to educe accurate and reliable knowledge. Systematic refinement of mind and body through the natural effortless process of transcending is said to result in subtler, more integrated experiences. These claims, which have immense positive implications for modern science and society, are said to be directly testable, supported by published research and open for anyone to investigate them using systematic experimental and experiential means.

Discussions about the nature of consciousness frequently involve the mind-body problem. From a developmental perspective, however, the mind-body problem confounds mind and consciousness. It is best associated with how gross relative and subtle relative levels of nature link up with each other, the theme of this book. In this sense the mind-body problem has little to do with what is and what is not conscious—neither of them being conscious *as such*. All mental activity of sensing, reasoning, and intuiting needs to be transcended in order to go beyond the mind-matter, subject-object duality of ordinary waking to the direct experience of pure consciousness itself.

This can be viewed as the key point of the Sankhya system, which is particularly relevant to progress in modern science toward understanding the full range of ontological levels of nature. In the Vedic tradition, this recognition leads to direct empirical validation beyond ordinary waking experience in the expanded epistemology that is the purpose of the Darshana of Yoga. Again, however, it is important to place Sankhya into its larger developmental framework of the other Darshanas. The Darshanas can be understood to identify the different phenomenal views of nature as one develops through higher states of consciousness to ultimate unity.

The descriptions of consciousness, mind and body in this book are primarily within the perspective of Sankhya. The conceptual recognition of consciousness as separate from mind and matter in Sankhya, and the direct experience of it in Yoga, are held to be the necessary basis for living the ultimate unity of Vedanta, the 'end of the Veda.'

Recent research on the structure of the cosmos in Vedic literature and locating its corresponding structure and function in human physiology is an unprecedented and truly revolutionary scientific advance toward validation of the holistic unity of matter, mind, and consciousness. [44, 90] This research is showing the infinitely self-interacting dynamical patterns of Veda to be reflected in the finite psychological, physiological, and behavioral levels of nature. When the full significance of this research is understood, it represents profound support for the completely holistic view of levels of nature, succinctly stated by Maharishi [91] that:

"The individual is cosmic."

In the completely holistic perspective of Vedanta, consciousness as universal Being is the reality underlying and permeating all relative phenomenal levels of nature. It is said to be direct empirical validation of the unified field as that which includes all relative subtle and gross levels of phenomenal existence. It is the direct experience that unifies nonlocal mind and local matter.

References

[1] Descartes R. (1972). *The philosophical works of Descartes* (2 vols.). Cambridge, England: Cambridge University Press.

[2] Locke J. (1959). *An essay concerning human understanding*. Fraser, A C. (Ed.). New York: Dover Publications. (Original work published 1690).

[3] Hagelin J. (2010). Foundations of physics and consciousness. Fairfield IA: Maharishi University of Management.

[4] Silberstein M. & McGeever J. (1999). The search for ontological emergence. *The Philosophical Quarterly, Vol. 49, No. 195*, 183-200.

[5] Boyer RW. (2008). *Bridge to unity: unified field-based science and spirituality*. Malibu: Institute for Advanced Research.

[6] Herbert N. (1985). *Quantum reality: beyond the new physics*. New York: Anchor Books.

[7] von Neumann, J. (1932). *Mathematische Grundlagen der Quantenmechanik*. Berlin: Springer Verlag. Trans. R. T. Beyer (1955). *Mathematical foundations of quantum mechanics*. Princeton: Princeton University Press.

[8] Farwell L. (2000). *How consciousness commands matter: The new scientific revolution and the evidence that anything is possible*. Iowa: Sunstar Publishing.

[9] Everett H. (1957). "Relative state" formulation of quantum mechanics. *Review of Modern Physics*, 29, July, 454-462.

[10] Greene B. (2004). *The fabric of the cosmos: Space, time, and the texture of reality*. New York: Alfred A. Knopf

[11] Einstein A., Podolsky B., & Rosen N. (1935). Can quantum-mechanical description of physical reality be considered complete? *Phys. Rev. 47*, 777-780.

[12] Clauser J., Horn M., Shimony A., & Holt R. (1969). *Phys. Rev. Lett., 26*, 880-884.

[13] Aspect A., Grangier P. & Roger G. (1982). Experimental Realization of Einstein-Podolsky-Rosen-Bohm Gedankenexperiment: A New Violation of Bell's Inequalities. *Phys. Rev. Lett.* 49, 91.

[14] Aspect A., Dalibard J. & Roger G. (1982). Experimental Test of Bell's Inequalities Using Time-Varying Analyzers. *Phys. Rev. Lett.* 49, 1804.

[15] Bell JS. (1964). On the Einstein-Podolsky-Rosen Paradox. *Physics*: 195-200.

[16] Bell JS. (1987). *Speakable and unspeakable in quantum mechanics.* Cambridge: Cambridge University Press.

[17] Bohm D. (1980). *Wholeness and the implicate order.* London: Routledge & Kegan Paul.

[18] Bohm D. & Hiley BJ. (1993). *The undivided universe.* London: Routledge.

[19] Dickson MW. (1998). *Quantum chance and non-locality.* Cambridge: Cambridge University Press.

[20] Radin D. & Borges A. (2009). *Explore: The Journal of Science and Healing,* 5, 4, July, 200-211.

[21] Greene B. (1999). *The elegant universe: superstrings, hidden dimensions, and the quest for the ultimate theory.* New York: Vintage Books.

[22] T'Hooft G. (2012). Quote from <http://worldsciencefestival.com/videos/a thin sheet of reality the universe as a hologram>

[23] Smolin L. (2001). *Three roads to quantum gravity.* New York: Basic Books.

[44] Wheeler JA. (1990). Information, physics, quantum: the search for links. In Zurek W. (Ed.). *Complexity, entropy, and the physics of information.* Redwood City, CA: Addison-Wesley.

[25] Shannon CE & Weaver W. *The mathematical theory of communication.* University of Illinois Press, Urbana, 1949.

[26] Verlinde E. (2010). *On the origin of gravity and the laws of Newton.* arXiv: 1001.0785v1[hep-th] 6 Jan.

[27] Nisargadatta Maharaj (1973). *I am That.* Durham, NC: Acorn Press.

[28] Woit P. (2006). *Not even wrong: The failure of string theory and the search for unity in physical law.* New York: Basic Books.

[29] Smolin L .(2006). *The trouble with physics: the rise of string theory, the fall of a science, and what comes next.* New York: Houghton Mifflin Co.

[30] Lincoln D. (2004). *Understanding the universe: from quarks to the cosmos.* Singapore: World Scientific Publishing Co., Pte. Ltd.

[31] Steinhardt P. (2003) The cyclic universe. In Brockman, J. (Ed.). *The new humanists: science at the edge.* New York: Barnes & Noble Books, pp. 297-311.

[32] Steinhardt PJ. (2011). The inflation debate: is the theory at the heart of modern cosmology deeply flawed? *Scientific American,* April, 38-43.

[33] Boyer RW. (2010). *Think outside the bang: beyond quantum theory and hidden dimensions to a holistic account of consciousness, mind and matter.* Malibu CA: Institute for Advanced Research.

[34] Zyga L. (2012, Jan 31). *Repulsive gravity as an alternative to dark energy (Part 1: In voids).* (Interview with Massimo Villata).

<physorg/news/2012-01-repulsive-gravity-alternative-dark-energy 1.html>

[35] Zyga L. (2012, Feb 1). *Repulsive gravity as an alternative to dark energy (Part 1: In the quantum vacuum).* (Interview with Dragan Hajdukovic). <physorg/news/2012-01-repulsive-gravity-alternative-dark-energy 1.html>

[36] Krauss LM. (2012). *A universe from nothing.* Free Press.

[37] Penrose R. (2005). *The road to reality. A complete guide to the laws of the universe.* New York: Alfred A. Knopf. (2005, pp. 17-23)

[38] Stapp HP. The Hard Problem: A Quantum Approach. In Shear, J. (Ed.) (2000). *Explaining consciousness—the hard problem.* Cambridge, MA: The MIT Press, pp. 197-215.

[39] Stapp HP. (2007). *Mindful universe: quantum mechanics and the participating observer.* Berlin Heidelberg New York: Springer-Verlag.

[40] Hagelin JS. (1989). Restructuring physics from its foundation in light of Maharishi's Vedic Science. *Modern Science and Vedic Science,* 3, 1, 3-72.

[41] *Maharishi Vedic University: Introduction.* (1994). Holland: Maharishi Vedic University Press.

[42] Maharishi Mahesh Yogi (1967). *Maharishi Mahesh Yogi on the Bhagavad-Gita: a new translation and commentary, chapters 1 to 6* (London: Penguin Books).

[43] Maharishi Mahesh Yogi (2004). Maharishi's Global News Conference, June 16.

[44] Nader T. (2000). *Human physiology: expression of Veda and the Vedic literature.* The Netherlands: Maharishi Vedic University.

[45] Bhavasar SN & Boyer RW. (2009). Major progress linking modern science and Vedic science. *Sambodhi,* Vol. XXXII, 1-32.

[46] Einstein A. (2006) as quoted in Albert Einstein home page, URL = <www.humboldt1.com/~gralsto/einstein/quotes.html>

[47] Maharishi Mahesh Yogi (2004). Maharishi's Global News Conference, June 23.

[48] Bernard T. (1947). *Hindu philosophy.* Delhi: Motilal Banarsidass Publishers.

[49] *Srimad Devi Bhagawatam* (1998). Swami Vijnananda (Tr.). New Delhi: Munshiram Manoharlal Publishers Pvt. Ltd, p. 145.

[50] *Celebrating Perfection in Education: Dawn of Total Knowledge.* (1997). India: Age of Enlightenment Publications (Printers).

[51] Radhakrishnan S. (1978). *The principle Upanishads*. USA: Humanities Press International (First published 1953 by George Allen and Unwin, Ltd.).

[52] Hagelin JS. (1987). Is consciousness the unified field? A field theorist's perspective. *Modern Science and Vedic Science*, 1, 1, January, 29-87.

[53] Varela FJ., Thompson E. & Rosch E. The embodied mind: cognitive science and human experience, Cambridge, MA. The MIT Press, 1993.

[54] Folger T. (2001). Quantum Shmantum. *Discover*, September.

[55] Maharishi Mahesh Yogi (2004). Global News Conference, March 17, MOU Channel.

[56] Huxley A. (1945). *The perennial philosophy*. New York: Harpers.

[57] Broadbent DE. (1958). *Perception and communication*. London: Pergamon Press.

[58] Shiffrin RM. & Schneider W. (1977). Controlled and automatic human information processing: II. Perceptual learning, automatic attending, and a general theory. Psychological Review, Vol. 84, No. 2, 127-189, 1977.

[59] Hameroff SR. (2008). The 'conscious pilot': synchronized dendritic webs move through brain neurocomputational networks to mediate consciousness, April 11 plenary session. Toward a Science of Consciousness Conference, April 8-12, 2008, Tucson, AZ.

[60] Wundt W. (1912). *Introduction to psychology*. London: Allen.

[61] Wundt W. (1907). *Outlines of psychology*. Leipzig: Wilhelm Engelman.

[62] Titchener EB. (1908). *Lectures on the elementary psychology of feeling and attention*. New York: The MacMillan Company.

[63] James W. (1890). *The principles of psychology*. New York: Holt.

[64] Baars B. J. (1997). *In the theatre of consciousness*. New York: Oxford University Press.

[65] Lazarus RS. (1984). On the primacy of cognition. *American Psychologist*, 39, 124-129.

[66] Zajonc RB. (1980). Feeling and thinking: preferences need no inferences. *American Psychologist*, 35, 151-175.

[67] Damasio A. (1999). *The feeling of what happens: Body and emotion in the making of consciousness*. New York: Harcourt, Inc.

[68] Boyer RW. (2011). The place and role of consciousness in human psychoarchitecture. *NeuroQuantology*, March-June-September-December, Vol. 9, Issue 1, 1-14.

[69] Sharma PV. (1981). (Ed., Trans.). *Caraka Samhita*, Vol. 1. Varanasi: Chaukhambha Orientalia.

[70] *Yoga Kundalini Upanishad* (2011). (Trans. K. Narayanasvami Aiyer). Vedic Scriptures Library on Astrojyoti.com. Retrieved 06 -03- 2011.

[71] Davies P. & Gregersen NH. (2010). Introduction: does information matter? In P. Davies & Gregersen NH. (Eds.) *Information and the nature of reality: from physics to metaphysics*. New York: Cambridge University Press; pp. 1-9.

[72] Dawkins R. (1989). *The selfish gene*. Oxford: Oxford University Press.

[73] Haney II W. S. (2006). *Cyberculture, cyborgs and science fiction: consciousness and the posthuman*. New York: Rodopi.

[74] Maharishi Mahesh Yogi (1963). *Science of being and art of living*. Washington, D.C.: Age of Enlightenment Publications.

[75] Maharishi Mahesh Yogi (1997). *Maharishi speaks to educators: mastery over Natural Law*, Vol. 4. India: Age of Enlightenment Publications (Printers).

[76] Maharishi Mahesh Yogi (1986). *Thirty years around the world—Dawn of the Age of Enlightenment*, I. The Netherlands: MVU Press.

[77] www.tm.org/research-on-meditation

[78] Orme-Johnson DW. (2010). www.truthabouttm.com

[79] *Scientific research on Maharishi's Transcendental Meditation and TM-Sidhi Programme—collected papers, Vols. 1-5 (1977-90)*. (Various Eds.) Fairfield, IA: Maharishi University of Management Press.

[80] Dillbeck MC. (2011). *Scientific research on Maharishi's Transcendental Meditation and TM-Sidhi Programme: collected papers, Vol. 6*. The Netherlands: Maharishi Vedic Univesity Press.

[81] Eppley KR, Abrams AI. & Shear J. (1989). Differential effects of relaxation techniques on trait anxiety: a meta-analysis. *Journal of Clinical Psychology*, 45, 957-974.

[82] Alexander CN., Rainforth MV. & Gelderloos P. (1991). Transcendental Meditation, self-actualization, and psychological health: a conceptual overview and statistical maet-analysis. *Journal of Social Behavior and Personality*, 6 (5): 189-247.

[83] Travis FT., Haaga DH., Hagelin JS., Tanner M., Arenander A, Nidich S., Gaylord-King C., Grosswald S, Rainforth M. & Schneider RH. (2009). A self-referential default brain state: patterns of coherence, power, and eLORETA sources during eyes-closed rest and the Transcendental Meditation practice. *Cognitive Processes*, 11 (1): 21-30.

[84] Barnes VA., Treiber FA. & Davis H. (2001). Impact of Transcendental Meditation on cardiovascular function at rest and during acute stress in

adolescents with high normal blood pressure. *Journal of Psychosomatic Research*, 51 4; D97-605.

[85] Brooks JS & Scarano T. (1985). Transcendental Meditation in the treatment of post-Vietnam adjustment. *Journal of Counseling and Development, 64; 212-215.*

[86] Orme-Johnson DW (1973). Autonomic stability and Transcendental Meditation. *Psychosomatic Medicine*, 35, 4; 341-349.

[87] Rosenthal NE (2011). *Transcendence: healing and transformation through Transcendental Meditation.* New York: Tarcher/Penguin.

[88] Travis F. & Arenander A. (2006). Cross-sectional and longitudinal study of effects of transcendental meditation practice on interhemispheric frontal asymmetry and frontal coherence. International Journal of Neuroscience,116; 1519-38.

[89] Travis F. & Shear J. (2010). Focused attention, open monitoring and automatic self-transcending: categories to organize meditations from Vedic, Buddhist and Chinese traditions. *Consciousness and Cognition,*19 (4): 1110-1118.

[90] Nader T. (2012). *Ramayan in human physiology, discovery of the eternal Reality of the Ramayan in the structure and function of human physiology.*

[91] Maharishi Mahesh Yogi (2003). Maharishi's Global News Conference, December 12.

Dedication & Acknowledgements

This book presents a holistic view that integrates ancient and modern knowledge for deeper appreciation of the vastness of nature and our place in and as it. It is a deeply fulfilling honor to dedicate the book to Maharishi Mahesh Yogi, who has been the greatest of teachers for knowledge of the integration of life. I hope the book accurately represents, at least to the degree I am capable, the profundity and revolutionary practical significance of his teaching of ultimate unity, for which I am forever grateful.

I first acknowledge the love and patience of my wife, Connie, and our wonderful families, each of whom brings unique blessings that support my work. I also want to acknowledge with deep appreciation the wisdom and practical feedback of Jerry and Debby Jarvis, Jonathan Shear who suggested I write on this topic, David Scharf, Fred Travis, David Orme-Johnson, and especially Park Hensley who brought a wealth of knowledge about physics without which this book would not have appeared on the surface of the Great Beyond.

Thanks also to you the reader. I hope you find the book worthy of your precious time, and would gratefully receive any feedback.

RW Boyer

www.ingramcontent.com/pod-product-compliance
Lightning Source LLC
Chambersburg PA
CBHW051508170526
45166CB00001B/441